ESD

ESD Series
By Steven H. Voldman

ESD: Analog Circuits and Design
ISBN: 9781119965183
August 2014

ESD Basics: From Semiconductor Manufacturing to Product Use
ISBN: 9780470979716
September 2012

ESD: Design and Synthesis
ISBN: 9780470685716
March 2011

ESD: Failure Mechanisms and Models
ISBN: 9780470511374
July 2009

Latchup
ISBN: 9780470016428
December 2007

ESD: RF Technology and Circuits
ISBN: 9780470847558
September 2006

ESD: Circuits and Devices
ISBN: 9780470847541
November 2005

ESD: Physics and Devices
ISBN: 9780470847534
September 2004

Upcoming titles:

ESD: Test and Characterization

Electrical Overstress (EOS): Devices, Circuits, and Systems

The ESD Handbook

ESD
Analog Circuits and Design

Steven H. Voldman
IEEE Fellow, Vermont, USA

WILEY

This edition first published 2015
© 2015 John Wiley & Sons, Ltd

Registered Office

John Wiley & Sons, Ltd, The Atrium, Southern Gate, Chichester, West Sussex, PO19 8SQ, United Kingdom

For details of our global editorial offices, for customer services and for information about how to apply for permission to reuse the copyright material in this book please see our website at www.wiley.com.

Library of Congress Cataloging-in-Publication Data

Voldman, Steven H.
  ESD : analog circuits and design / Steven H. Voldman.
    1 online resource.
  Includes bibliographical references and index.
  Description based on print version record and CIP data provided by publisher; resource not viewed.
  ISBN 978-1-118-70147-8 (ePub) – ISBN 978-1-118-70168-3 (Adobe PDF) – ISBN 978-1-119-96518-3 (cloth)   1. Semiconductors–Protection.   2. Analog integrated circuits–Protection.   3. Analog integrated circuits–Design and construction.   4. Electrostatics.   5. Static eliminators.   I. Title.
  TK7871.85
  621.3815′3–dc23

                                                            2014021528

A catalogue record for this book is available from the British Library.

ISBN: 978-1-119-96518-3

Set in 10/12pt Times New Roman by SPi Publisher Services, Pondicherry, India
Printed and bound in Malaysia by Vivar Printing Sdn Bhd

1   2015

Contents

About the Author

Dr. Steven H. Voldman is the first IEEE Fellow in the field of electrostatic discharge (ESD) for "Contributions in ESD protection in CMOS, Silicon on Insulator and Silicon Germanium Technology." He received his B.S. in Engineering Science from the University of Buffalo (1979), a first M.S. EE (1981) from Massachusetts Institute of Technology (MIT), a second EE Degree (Engineer Degree) from MIT, a M.S. Engineering Physics (1986), and a Ph.D. in Electrical Engineering (EE) (1991) from University of Vermont under IBM's Resident Study Fellow program.

Voldman was a member of the semiconductor development of IBM for 25 years. He was a member of the IBM's Bipolar SRAM, CMOS DRAM, CMOS logic, Silicon on Insulator (SOI), 3-D memory team, BiCMOS and Silicon Germanium, RF CMOS, RF SOI, smart power technology development, and image processing technology teams. In 2007, Voldman joined the Qimonda Corporation as a member of the DRAM development team, working on 70, 58, 48, and 32 nm CMOS DRAM technology. In 2008, Voldman worked as a full-time ESD consultant for Taiwan Semiconductor Manufacturing Corporation (TSMC) supporting ESD and latchup development for 45 nm CMOS technology and a member of the TSMC Standard Cell Development team in Hsinchu, Taiwan. From 2009 to 2011, Steve was a Senior Principal Engineer working for the Intersil Corporation working on analog, power, and RF applications in RF CMOS, RF Silicon Germanium, and SOI. From 2013 to 2014, Dr. Voldman was a consultant for the Samsung Electronics Corporation in Dongtan, South Korea.

Dr. Voldman was the chairman of the SEMATECH ESD Working Group, from 1995 to 2000. In his SEMATECH Working Group, the effort focused on ESD technology benchmarking, the first transmission line pulse (TLP) standard development team, strategic planning, and JEDEC-ESD Association standards harmonization of the human body model (HBM) Standard. From 2000 to 2013, as Chairman of the ESD Association Work Group on TLP and very-fast TLP (VF-TLP), his team was responsible for initiating the first standard practice and standards for TLP and VF-TLP. Steve Voldman has been a member of the ESD Association Board of Directors, and Education Committee. He initiated the "ESD on Campus" program which was established to bring ESD lectures and interaction to university faculty and students internationally; the ESD on Campus program has reached over 40 universities in the United States, Korea, Singapore, Taiwan,

Malaysia, Philippines, Thailand, India, and China. Dr. Voldman has taught short courses and tutorials on ESD, latchup, patenting, and invention.

He is a recipient of 250 issued US patents and has written over 150 technical papers in the area of ESD and CMOS latchup. Since 2007, he has served as an expert witness in patent litigation and has also founded a limited liability corporation (LLC) consulting business supporting patents, patent writing, and patent litigation. In his LLC, Voldman served as an expert witness for cases on DRAM development, semiconductor development, integrated circuits, and electrostatic discharge. He is presently writing patents for law firms. Steven Voldman provides tutorials and lectures on inventions, innovations, and patents in Malaysia, Sri Lanka, and the United States.

Dr. Voldman has also written many articles for *Scientific American* and is an author of the first book series on ESD and latchup (eight books): *ESD: Physics and Devices, ESD: Circuits and Devices, ESD: Radio Frequency (RF) Technology and Circuits, Latchup, ESD: Failure Mechanisms and Models, ESD Design and Synthesis, ESD Basics: From Semiconductor Manufacturing to Product Use,* and *Electrical Overstress (EOS): Devices, Circuits and Systems,* as well as a contributor to the book *Silicon Germanium: Technology, Modeling and Design* and *Nanoelectronics: Nanowires, Molecular Electronics, and Nano-devices.* In addition, the International Chinese editions of book *ESD: Circuits and Devices, ESD: Radio Frequency (RF) Technology and Circuits,* and *ESD Design and Synthesis* (2014) are also released.

Preface

This book *Electrostatic Discharge (ESD): Analog Circuits and Design* was initiated based on the need to produce a text that addresses the fundamentals of electrostatic discharge (ESD) requirements for analog and power electronic devices, components and systems. As the manufacturing world evolves, semiconductor devices are scaling and systems are changing. As a result, the needs and requirements of reliability and ESD robust products are increasing. A text is required that connects and synthesizes the fundamentals of analog design discipline and the ESD discipline. Whereas significant texts are available today to teach experts on ESD on-chip design for digital design discipline and radio frequency (RF) design discipline, there is no single textbook devoted to ESD on-chip design dedicated to analog design.

With the growth of mixed signal applications that integrate both the digital and analog circuitries on a common semiconductor chip, there is a growing interest in the concerns associated with ESD design and protection within a digital–analog application. New issues arise by the integration of two separate cores within a chip, where the digital and the analog domains are separated spatially and electrically. As a result, new problems arise with the design synthesis and architecture of an analog–digital mixed signal semiconductor chip.

With the growth of power management and power devices, there is an additional challenge in providing ESD protection within a power application. In power technology, the number of allowed power conditions has grown significantly, leading to a significant difficulty to provide ESD protection.

In addition, there is a growing interest in electrical overstress (EOS). Today, there is a need for understanding the fundamentals of EOS. Necessarily experts, non-experts, non-technical staff and layman should understand the problems that the world faces today. Today, real-world EOS issues surround us; this occurs in manufacturing environments, power sources, machinery, actuators, solenoids, soldering irons, cables, and lightning. When there is switching, poor grounding, ground loops, noise and transient phenomena, there is a potential for EOS of devices, components, and printed circuit boards. Hence, there is a need for experts and non-experts to understand the issues that revolve around us and the steps to be taken to avoid them. At present, this book is the only textbook on the issues of EOS. In this book, EOS issues for analog and power applications are emphasized.

Hence, there is an opportunity to intermix the analog design discipline and the ESD design discipline to produce a synthesized "ESD analog design discipline" that utilizes analog design techniques and ESD protection techniques.

This book has multiple goals and are as follows.

The first goal is to teach the basic and fundamental concepts of the analog design discipline.

The second goal is to review the analog circuit building blocks used in analog design, such as current mirrors, error amplifiers, feedback loops, and comparators, and to discuss the bandgap reference circuits and low dropout regulators.

The third goal is to discuss EOS and ESD, that is to explain the practices, devices, and novel concepts to address both ESD and EOS on-chip and off-chip.

The fourth goal is to discuss the needs for analog circuits and design associated with CMOS latchup. This involves understanding of latchup in a mixed signal digital–analog environment.

The fifth goal is to explain the semiconductor chip floorplanning to address ESD, EOS and latchup in an analog–digital mixed signal environment. Latchup issues, placement, and guard ring requirements will be highlighted.

The sixth goal is to describe the novel concepts that provide both analog and ESD advantages.

And, the last goal is to highlight the electrical design automation (EDA) methods for analog, analog–digital mixed signal, and power electronics to provide ESD and EOS robust products.

This book is organized as to allow the reader to learn the fundamentals of ESD analog design.

In Chapter 1, analog design principles associated with matching and design symmetry are discussed, and EOS and its relationship with other phenomena, such as ESD, electromagnetic interference, electromagnetic compatibility, and latchup are explained. EOS is defined as well in terms of electrical over-current, electrical over-power, and other concepts. ESD and EOS events on analog applications are also emphasized. As a result, we will draw distinctions through the text on difference of failure analysis, time constants, and other means of identification and classification. A plan to define safe operating area and its role in EOS is also emphasized.

In Chapter 2, the analog design layout practices of interdigitated design layout and common centroid concepts in one- and two dimensions are discussed. These concepts are implemented into ESD networks and the cosynthesis of analog circuits and ESD networks.

In Chapter 3, examples of analog building blocks and circuits that exist in analog designs are provided for readers unfamiliar with analog circuit networks. The analog circuit examples include single-ended receivers, differential receivers, comparators, current mirrors, bandgap regulators, and voltage converters. Voltage regulators of interest include buck, boost, buck–boost, and other circuit topologies.

In Chapter 4, the analog ESD design discipline is introduced, applying both the ESD requirements and the layout concepts of analog circuitry. This includes the digital ESD design discipline, in contrast to the analog ESD design discipline.

In Chapter 5, the analog design synthesis on a high level, by addressing the floorplanning of a mixed signal chip application is discussed. This includes the digital and analog power domain floorplanning, digital and analog power grid, digital to analog breaker cells, and ESD concerns in digital–analog mixed signal chips. Additionally, the guard ring and moats within the chip architecture are discussed, and active and passive guard ring concepts are shown.

In Chapter 6, the signal line ESD failures in digital and analog domains where they are required to be decoupled due to noise are addressed. ESD solutions between the ground connections include coupling using resistors, diode elements, as well as third-party functional blocks. ESD solutions along the signal lines include the ground connections as well as the ESD networks on the signal lines that cross the digital to analog domain.

In Chapter 7, the analog and ESD signal pin cosynthesis that introduces the usage of inter-digitated layout and common centroid concepts to provide ideal matching, low capacitance, and small area in differential signal pins is addressed. Using interdigitated designs, the parasitic elements are utilized for signal pin-to-signal pin ESD protection. In conclusion, the issue of common centroid design of ESD protection networks which integrates signal pin-to-signal pin ESD protection with the inter-digitation pattern for ESD pin-to-rail protection network for differential pair circuitry is discussed for the first time. With integration of the ESD pin-to-rail solution and the ESD signal pin solution, a significant reduction in the area and loading effect for CMOS differential circuits is established. This novel concept will provide significant advantage for present and future high-performance analog and RF design for matching, area reduction, and performance advantages.

In Chapter 8, analog and ESD integration is focused. Topics such as analog signal pin input circuitry and ESD power clamps are discussed, and ESD power clamp issues and solutions are explained in more details. Significant discussion is provided due to the importance of ESD power clamps in analog and digital ESD design.

In Chapter 9, the system-level issues associated with EOS in chips, printed circuit boards, and systems are discussed. EOS protection device classifications, symbols, and types for both over-voltage or over-current conditions are highlighted. System-level and system-like testing methods, such as IEC 61000-4-2, IEC 61000-4-5, and human metal model waveforms and methods, are reviewed. Examples for printed circuit board design for digital–analog systems are also provided. The EDA techniques and methods for ESD in analog design are also discussed. Methods such as design rule checking (DRC), layout versus schematic (LVS), and electrical rule checking (ERC) are used for both ESD and EOS checking and verification.

In Chapter 10, latchup in analog design is discussed and solutions to avoid digital noise from impacting analog circuitry are addressed. The spatial placement of digital and analog cores in a mixed signal chip as well as the guard rings between these domains are also explained. Moats, guard rings, and through-silicon via advantages and disadvantages as possible solutions to minimize both noise and latchup are highlighted. Special features, such as grounded wells, and decoupling capacitor issues and how they can lead to latchup in analog applications are also reviewed. In conclusion, I/O to I/O interactions as a function of standard cell-to-standard cell spacings are discussed. As technology spacings are reduced, cell-to-cell latchup will increase in importance in analog design.

In Chapter 11, ESD and EOS libraries and documents for an analog or mixed signal technology are discussed. The discussion includes a plethora of items, from analog libraries, ESD library elements, Cadence™-based parameterized cells, Cadence-based hierarchical ESD designs, to ESD cookbooks. ESD and EOS documents for technology design manual, cookbooks, checklists, and design release processes are discussed. Control programs and documents for analog ESD design are highlighted. This includes ESD analog design "cookbooks" to assist design teams to determine the correct elements to use for a given circuit.

In Chapter 12, EDA techniques and methods for ESD, EOS, and latchup are discussed. Methods such as DRC, LVS and ERC are used for ESD, latchup, and EOS checking and verification. As time progresses, ESD CAD methods are being propagated to EOS CAD methods, to address ESD and EOS in the same design tool. The example of Calibre PERC™ shows how the methods of ESD are being extended to the EOS issue. A key issue is the checking and verification of analog-to-digital cross domain sign lines. This trend will continue in the future.

This introductory text will hopefully open your interest in the field of ESD in analog design. To establish a stronger knowledge of ESD protection, it is advisable to read other texts such as *ESD Basics: From Semiconductor Manufacturing to Product Use, ESD: Physics and Devices, ESD: Circuits and Technology, ESD: RF Circuits and Technology, ESD: Failure Mechanisms and Models, ESD: Design and Synthesis, EOS: Devices, Circuits and Systems,* and *Latchup.*

Enjoy the text, and enjoy the subject of ESD design in analog devices, circuitry, and systems.

Baruch HaShem B"H

Dr. Steven H. Voldman
IEEE Fellow

Acknowledgments

I would like to thank the years of support from the Electrostatic Discharge (ESD) Association, the Taiwan Electrostatic Discharge (T-ESD) Association, the IEEE, and the JEDEC organizations. I would like to thank the IBM Corporation, Qimonda, Taiwan Semiconductor Manufacturing Corporation (TSMC), Intersil Corporation, and Samsung Electronics Corporation. I was fortunate to work in a wide number of technology teams and with a wide breadth of customers. I was very fortunate to be a member of talented technology and design teams that were both innovative, intelligent, and inventive.

I would like to thank the institutions that allowed me to teach and lecture at conferences, symposiums, industry, and universities; this gave me the motivation to develop the texts. I would like to thank faculty at the following universities: M.I.T., Stanford University, University of Central Florida (UCF), University Illinois Urbana-Champaign (UIUC), University of California Riverside (UCR), University of Buffalo, National Chiao Tung University (NCTU), Tsin Hua University, National Technical University of Science and Technology (NTUST), National University of Singapore (NUS), Nanyang Technical University (NTU), Beijing University, Fudan University, Shanghai Jiao Tung University, Zheijang University, Huazhong University of Science and Technology (HUST), UESTC, Korea University, Universiti Sains Malaysia (USM), Universiti Putra Malaysia (UPM), Kolej Damansara Utama (KDU), Chulalongkorn University, Mahanakorn University, Kasetsart University, Thammasat University, and Mapua Institute of Technology (MIT).

I would like to thank for the years of support and the opportunity to provide lectures, invited talks, and tutorials the *Electrical Overstress/Electrostatic Discharge (EOS/ESD) Symposium*, the *International Reliability Physics Symposium (IRPS)*, the *Taiwan Electrostatic Discharge Conference (T-ESDC)*, the *International Electron Device Meeting (IEDM)*, the *International Conference on Solid-State and Integrated Circuit Technology (ICSICT)*, the *International Physical and Failure Analysis (IPFA)*, *IEEE ASICON*, and *IEEE Intelligent Signal Processing And Communication Systems (ISPACS) Conference.*

I would like to thank my many friends for over 25 years in the ESD profession: Prof. Ming Dou Ker, Prof. J.J. Liou, Prof. Albert Wang, Prof. Elyse Rosenbaum, Timothy J. Maloney, Charvaka Duvvury, Eugene Worley, Robert Ashton, Yehuda Smooha, Vladislav Vashchenko, Ann Concannon, Albert Wallash, Vessilin Vassilev, Warren

Anderson, Marie Denison, Alan Righter, Andrew Olney, Bruce Atwood, Jon Barth, Evan Grund, David Bennett, Tom Meuse, Michael Hopkins, Yoon Huh, Jin Min, Keichi Hasegawa, Teruo Suzuki, Nathan Peachey, Kathy Muhonen, Augusto Tazzoli, Gaudenzio Menneghesso, Marise BaFleur, Jeremy Smith, Nisha Ram, Swee K. Lau, Tom Diep, Lifang Lou, Stephen Beebe, Michael Chaine, Pee Ya Tan, Theo Smedes, Markus Mergens, Christian Russ, Harold Gossner, Wolfgang Stadler, Ming Hsiang Song, J.C. Tseng, J.H. Lee, Michael Wu, Erin Liao, Stephen Gaul, Jean-Michel Tschann, Han-Gu Kim, Kitae Lee, Ko Noguchi, Tze Wee Chen, Shu Qing Cao, Slavica Malobabic, David Ellis, Blerina Aliaj, Lin Lin, David Swenson, Donn Bellmore, Ed Chase, Doug Smith, W. Greason, Stephen Halperin, Tom Albano, Ted Dangelmayer, Terry Welsher, John Kinnear, and Ron Gibson.

I would like to thank the ESD Association office for the support in the area of publications, standards developments, and conference activities. I would also like to thank the publisher and staff of John Wiley & Sons, for including this text as part of the ESD book series.

To my children, Aaron Samuel Voldman and Rachel Pesha Voldman, good luck to both of you in the future.

To my wife Annie Brown Voldman, of 30 years of love and friendship—thank you for your patience and support.

And to my parents, Carl and Blossom Voldman.

Baruch HaShem B"H

Dr. Steven H. Voldman
IEEE Fellow

1 Analog, ESD, and EOS

In 1993, I was invited to consult for two days with a well-known semiconductor analog corporation on electrostatic discharge (ESD) protection of analog components. A vice president of the corporation sat with me and said, "Our analog products are superior to any of our competitors. As a result, no one cared about the level of our ESD protection results! All of our products did not achieve 2000 V HBM or 4000 V HBM levels. Today, with growth of competition in the analog business sector, overnight, 75% of our customers want us to achieve better than 2000 V HBM levels on all of our products! How do I build a corporate ESD strategy for this analog corporation ?..."

This was my first introduction to the world of ESD in analog design.

1.1 ESD IN ANALOG DESIGN

In every technology sector, electrostatic discharge (ESD) protection was not an issue when there was a sole supplier of critical products and the customer was willing to accept the product. Eventually, as the technology or application space matured, customers wanted better ESD protection as both technology and application became mainstream or high volume. This was true historically in digital and analog applications with CMOS, bipolar, silicon on insulator (SOI), silicon germanium (SiGe), and gallium arsenide (GaAs) technologies. With mainstream introduction of a technology, it is desirable not to have customer field returns from ESD or electrical overstress (EOS).

As a result of the unique needs of analog design, there are a significant number of issues to be addressed in analog ESD design. These issues extend from chip design to system-level design in both architecture and layout, which consist of the following:

- Matching and layout issues

- Matching requirements in differential receivers

ESD: Analog Circuits and Design, First Edition. Steven H. Voldman.
© 2015 John Wiley & Sons, Ltd. Published 2015 by John Wiley & Sons, Ltd.

- Domain-to-domain separation and ESD coupling
- Circuit topology chip architecture and ESD
- Interdomain digital-to-analog ESD failures
- Semiconductor chip layout floor planning
- Printed circuit board (PCB) design floor planning
- High-voltage applications
- Ultrahigh-voltage applications

In this text, examples will be provided of ESD failures and problems in past analog applications to modern-day practices in analog design. The text will discuss how the present-day architecture of mixed-signal chips evolved and its implications.

1.2 ANALOG DESIGN DISCIPLINE AND ESD CIRCUIT TECHNIQUES

In analog design, unique design practices are used to improve the functional characteristics of analog circuitry [1–10]. In the ESD design synthesis of analog circuitry, the ESD design practices must be suitable and consistent with the needs and requirements of analog circuitry [11, 12] (Figure 1.1). Fortunately, many of the analog design practices are aligned with ESD design practices.

In the analog design discipline, there are many design techniques to improve tolerance of analog circuits [10]. Analog design techniques include the following:

- Local matching: Placement of elements close together for improved tolerance
- Global matching: Placement in the semiconductor die
- Thermal symmetry: Design symmetry

A key analog circuit design requirement is matching. To avoid semiconductor process variations, matching is optimized by the local placement. Placement within the die location also is an analog concern due to mechanical stress effects. In analog design, there is a concern of the temperature field within the die and the effect of temperature distribution within the die.

Many of the analog design synthesis and practices are also good ESD design practices. The design practices of matching and design symmetry are also suitable practices for

Figure 1.1 Analog and ESD design.

ESD design. But there are some design practices where a trade-off exists between the analog tolerance and ESD; this occurs when parasitic devices are formed between the different analog elements within a given circuit or circuit to circuit.

1.2.1 Analog Design: Local Matching

Matching is important in analog design due to the usage of many circuit blocks that require good matching characteristics. The matching is important locally in a semiconductor device or within a circuit. In this case, "local matching" is needed to provide the ideal characteristics of an analog network. Local matching is critical in multifinger structures, where mismatch can occur between two adjacent structures. In future sections, discussion of semiconductor processes such as photolithography and etching influences the local matching.

1.2.2 Analog Design: Global Matching

Matching is important in analog design due to the usage of many circuit blocks that require good matching characteristics from circuit to circuit. Many functional analog circuit blocks are repeated within a semiconductor chip. In this case, "global matching" is needed to avoid mismatch between two circuits. Global matching is influenced by spatial separation of two circuits, global density variations, arrangement, and orientation. Global matching is influenced by across chip linewidth variation (ACLV).

1.2.3 Symmetry

Symmetry is critical to establish matching within a semiconductor device or an analog circuit (Figure 1.2). Symmetry is influenced by design layout, current distribution, temperature field, and thermal distribution [10].

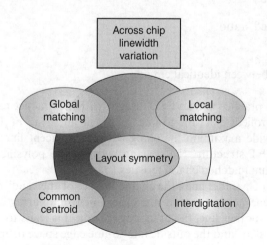

Figure 1.2   Symmetry and layout.

1.2.3.1 Layout Symmetry

Layout symmetry is a form of symmetry through physical design. Layout symmetry establishes matching within a semiconductor device or an analog circuit [10].

1.2.3.2 Thermal Symmetry

Thermal symmetry is a form of symmetry through the temperature field and thermal distribution. Thermal symmetry establishes matching within a semiconductor device or an analog circuit [10].

1.2.4 Analog Design: Across Chip Linewidth Variation

In semiconductor development, semiconductor process variation can introduce structural and dimensional nonuniformity which influences both analog circuits and ESD devices [13–22]. Photolithography and etch tools can introduce these nonuniformities that exist on a local and global design level. These variations can manifest themselves by introducing variations in both active and passive elements. For MOSFET transistors, variation in the MOSFET channel length in single-finger and multifinger MOSFET layouts can lead to nonuniform "turn-on"; this effect can influence both active functional circuits and ESD networks. In bipolar transistors, the linewidth variation can lead to different sizes in emitter structures, leading to nonuniform current distribution in multifinger bipolar transistors. For resistor elements, resistor elements that are utilized for ballasting in multifinger structures can also lead to nonuniform current in the different fingers in the structure.

Design factors that influence the lack of variation are the following semiconductor process and design variables:

- Linewidth

- Line-to-line space

- "Nested-to-isolated" ratio

- Orientation

- Physical spacing between identical circuits

It is a circuit design practice and an ESD design synthesis practice to provide a linewidth which is well controlled. For line-to-line space, in an array of lines, the spacing is maintained to provide maximum matching between adjacent lines. For example, in a multifinger MOSFET structure, the spacing between the polysilicon lines is equal to provide the maximum matched characteristics.

Given any array of parallel lines, the characteristics of the "end" or edges of the array can have different characteristics than the other lines. In an array of lines, whereas one edge is adjacent to another line, the other edge is not; this leads to one line-to-line edge space to appear "nested" and the outside line-to-line edge space to appear "semi-infinite" or "isolated." To address the problem of poorly matched edge lines, the following semiconductor process and ESD design solutions are used:

- Process: Cancellation technique of photolithography and etch biases
- Design: Use of dummy edge lines
- Circuit: Use of "gate-driven" circuitry

Orientation can also influence the linewidth of identical circuits both locally and globally. An ESD design practice is to maintain the same x–y orientation of ESD circuits in a semiconductor chip to minimize variation pin to pin. This is not always possible in a peripheral architecture where the ESD element is rotated on the four edges of the semiconductor chip. Note that in this case, the circuit itself (e.g., off-chip driver) may also undergo an orientation effect. It is a good ESD design synthesis practice that addresses the orientation issue with compensation and matching issues for orientation of the ESD elements (in conjunction with the circuit it is protecting).

On a macroscopic full-chip scale, variations in the photolithography and etching can vary from the top to bottom of a semiconductor chip. In the design of a semiconductor chip, these can be compensated with a preknowledge of the photolithography and etch variations of a technology.

1.3 DESIGN SYMMETRY AND ESD

Design symmetry is an ESD design practice to maximize the ESD robustness [23]. The capability of the ESD network to dissipate high-current pulse events is directly related to the network's topology and its design symmetry. The more uniform the current distribution is through the ESD network during a discharge, the better the utilization of the area of the structure, and as a consequence, the greater the robustness of the circuit design. The distribution of current during an ESD event is dependent upon the design symmetry of the ESD network and its components.

From experience, to the degree that the design of the ESD network (or structure) on all levels of the integrated circuit (IC) departs from a symmetric configuration, the greater is the current localized or nonuniformities in the ESD network. With a symmetrical distribution of the current, the peak power-to-failure per unit area is lowered, producing superior results. Additionally, the more uniform the current distribution, the more uniform the thermal field as well. Since semiconductor element electrical and thermal parameters are temperature dependent (e.g., mobility, electrical conductivity, thermal conductivity), the more uniform the current distribution, the more symmetrical the temperature distribution within the device.

In IC design, a key ESD design concept is to maintain a high degree of design symmetry within a structure on all design levels. In both the ESD network and I/O driver circuit, an evaluation of the power distribution of an ESD event within the circuit is an indicator of the robustness of the IC. Hence, physical layout design symmetry can be used as a heuristic determination of the power distribution within a physical structure.

To evaluate ESD design symmetry, this can be done visually in a design review or through a means of an automated computer-aided design (CAD) tool [23].

ICs are produced on a uniform substrate which are the subject of numerous mask operations. The masks create from lines and shapes individual devices on the layers of the ICs. Hence, the mask physical layout features can be used to quantify the ESD design symmetry. This can be done on each of the layout design levels of the ESD structure.

To define ESD design symmetry, an axis of symmetry can be defined in the ESD design. Semiconductor design layout is two dimensional, allowing to define an axis of symmetry in the x- and y-direction. In this fashion, "moments" can be defined about the axis of symmetry as a means of quantifying the degree of symmetry and identify non-symmetric features.

Before manufacturing the IC in silicon, the data file which defines the lines and shapes of each mask is available for evaluating the design to be implemented in silicon. In this methodology, a method can define the symmetry which evaluates on a level-by-level basis.

In this design methodology, the method provides for evaluating the degree of design symmetry of the proposed semiconductor device, by considering various topological features of the design such as the directional flow of current into and out of the device, circuit element design symmetry, metal and contact symmetry, and other design features which reduce the robustness of an ESD protection network.

The semiconductor design can be "checked" before implementing in silicon by evaluating each ESD shapes. In the event that any level fails the check, the level can be redesigned before implementing in silicon.

1.4 ESD DESIGN SYNTHESIS AND ARCHITECTURE FLOW

In the ESD design synthesis process, there is a flow of steps and procedures to construct a semiconductor chip. The following design synthesis procedure is an example of an ESD design flow needed for semiconductor chip implementations [13, 14, 22]:

- **I/O, domains, and core floor plan:** Define floor plan of regions of cores, domains, and peripheral I/O circuitry

- **I/O floor plan:** Define area and placement for I/O circuitry

- **ESD signal pin floor plan:** Define ESD area and placement

- **ESD power clamp network floor plan:** Define ESD power clamp area and placement for a given domain

- **ESD domain-to-domain network floor plan:** Define ESD networks between the different chip domains area and placement for a given domain

- **ESD signal pin network definition:** Define ESD network for the I/O circuitry

- **ESD power clamp network definition:** Define ESD power clamp network within a power domain

- **Power bus definition and placement:** Define placement, bus width, and resistance requirements for the power bus

- **Ground bus definition and placement:** Define placement, bus width, and resistance requirements for the ground bus

- **I/O to ESD guard rings:** Define guard rings between I/O and ESD networks

- **I/O internal guard rings:** Define guard rings within the I/O circuitry

- **I/O external guard rings:** Define guard rings between I/O circuitry and adjacent external circuitry.

1.5 ESD DESIGN AND NOISE

In semiconductor chips, the switching of circuitry can lead to noise generation that can influence circuit functions [17]. In the architecture of a semiconductor chip, different domains are separated due to noise generation. Noise is generated from undershoot and overshoot phenomena of signal leading to substrate injection. Noise can be also generated from switching of off-chip drivers circuitry and large power devices.

In a semiconductor chip, there are circuitries which are sensitive to the noise generation. To avoid noise from impacting functionality of the sensitive circuitry, separate power domains are created on a common substrate. Examples of separation of power domains and semiconductor chip functions are as follows [13, 22] (Figure 1.3):

- Separation of peripheral circuitry from core circuitry

- Separation of peripheral circuitry from core memory regions

- Separation of digital and analog functions

- Separation of digital, analog, and radio frequency (RF) functions

- Separation of power, digital, and analog functions

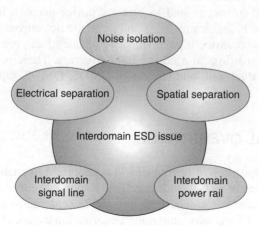

Figure 1.3   Noise and spatial separation.

In ESD design, the separation of different circuit domains for noise isolation can lead to ESD failures. ESD failures can occur due to the following situations:

- Lack of a forward bias current path between a first power rail and a second power rail
- Lack of a forward bias current path between a pin of a first domain and a power rail of a second domain
- Lack of a forward bias current path between a pin of a first domain and a pin of a second domain

As a result, in the ESD design synthesis, the architecture of the semiconductor chip which is separated for noise must allow for current to flow from domain to domain. As a result, there is a trade-off between the noise isolation and the requirement of allowance of current flow between pins and power rails of different domains. Various means have been utilized to achieve this. Examples of solutions between independent power domains are as follows:

- Symmetric and asymmetric bidirectional ESD networks between domains of $V_{DD}(i)$ and $V_{DD}(j)$
- Symmetric and asymmetric bidirectional ESD networks between domains of $V_{DD}(i)$ and $V_{SS}(j)$
- Symmetric and asymmetric bidirectional ESD networks between domains of $V_{SS}(i)$ and $V_{SS}(j)$

1.6 ESD DESIGN CONCEPTS: ADJACENCY

An ESD design concept is the issue of adjacency. In the physical layout design of ESD structures, the adjacency of structures internal and external of the ESD element is a concern.

Structures adjacent to ESD structures can lead to both ESD failure mechanisms and latchup. Adjacent structures can lead to parasitic device elements not contained within the circuit schematics. Parasitic npn, pnp, and pnpn are not uncommon in the design of ESD structures. These parasitic bipolar transistor elements can occur between the ESD structure and the guard rings around the ESD structure. These parasitic elements can occur in single-well, dual-well, and triple-well CMOS, DMOS, bipolar, BiCMOS, and BCD technologies.

1.7 ELECTRICAL OVERSTRESS

EOS is such a broad spectrum of phenomena; it is important to establish classifications of EOS [13]. The definition of EOS includes electrical response to current, voltage, and power.

Electrical phenomena are categorized into different definitions, which will be discussed in depth in further sections. Common categorization includes ESD, electromagnetic interference (EMI), electromagnetic compatibility (EMC), and latchup issues (Figure 1.4)

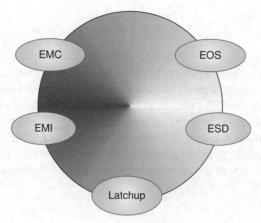

Figure 1.4   EOS, ESD, EMI, EMC, and latchup.

[13, 14, 22]. At times, all of these are included in the definition of EOS; yet others separate these categories as separate items to distinguish them for the purpose of determining cause–effect relationships, as well as root cause. For example, although ESD is a form of EOS, it is established in the semiconductor industry to distinguish between ESD and EOS. One of the reasons this is done is due to determining the root cause of failure.

EOS cause and effect for ICs can be the following [13]:

- ESD
- Latchup
- EMI
- EMC
- Misapplication

For ESD phenomena, there exist event models for the component and system levels. For component-level ESD, failures can be associated with human body model (HBM), machine model (MM), charged device model (CDM), and human metal model (HMM) [13–22]. For system-level ESD, failures can be associated with charged board model (CBM) and cable discharge event (CDE) [13, 14].

For latchup, there exist causes associated with direct current (DC) and transient phenomena [18]. DC latchup events can be in the form of "internal latchup" and "external latchup." Transient latchup is also the initiation of latchup from a transient voltage event.

For EMI, EOS events can occur from the following [13]:

- Noise
- Surge currents
- Slow voltage transients

- Fast voltage transients

- RF signals

For the EMI events, there are causes for noise, surge currents, transients, and RF interferences. Noise can be a result of lack of proper filters and switching events. Surge currents can occur due to poor electrical isolation and switching of capacitors. Voltage transients can occur due to power-up and power-down of PCB and ICs. Inductive switching is also a transient voltage concern. RF interference can be a concern from lack of filters, lack of shielding, shielding openings, and the PCB design quality [13, 14, 17].

Human error and misapplication is a large cause of EOS events. This can happen in the following forms:

- System design

- Improper testing

- Improper assembly

- Specification violation

EOS can be a result of poor system design [13]. System design can be both hardware and software. Improper or inadequate design of both the electrical and thermal properties can lead to EOS.

EOS events can be the result of improper testing [13]. Human error from incomplete tests, hot swapping, and switching of components, and inadequate margins can lead to over-voltage and over-current of applications. Overvoltage can also occur in the test equipment sources themselves due to noise, transient spikes, and other poor-quality test environments.

Improper assembly and human error can also be the cause of EOS issues. In the assembly process, misorientation, misinsertion, reverse insertion, and assembly of powered or unpowered states can lead to EOS.

In addition, electrical specifications can be violated due to defective hardware (e.g., opens and shorts), poor electrical contacts, poor ground connections, and overheating.

Throughout the text, these issues will be reemphasized, repeated, and addressed in detail. To continue with our discussion, more definitions will be established in this chapter.

1.7.1 Electrical Overcurrent

There are different forms of EOS [13]. In electrical conditions that are in excess of the intended or application current, devices, components, or systems can undergo latent or permanent damage; this condition can be defined as electrical overcurrent (EOC).

When EOC occurs, electronic components can have excessive Joule heating, material property changes, melting, or fire. EOC is one classification of EOS. EOC can be prevented by electrical fuses, temperature sensing circuitry, and current-limiting EOS protection devices.

1.7.2 Electrical Overvoltage

In electrical conditions that are in excess of the intended or application voltage, devices, components, or systems can undergo latent or permanent damage; this condition can be defined as electrical overvoltage (EOV). When EOV occurs, electronic components can undergo different conditions. EOV can lead to electrical breakdown of dielectrics, semiconductors, and conductors and is a second classification of EOS.

EOV can be prevented by voltage-limiting EOS protection devices and ESD protection circuits.

1.7.3 Electrical Overstress Events

EOS can occur within manufacturing environments and production areas and in the field [13]. EOS events can occur internal or external of electronic systems. External sources can be associated with voltage sources, current sources, and phenomena associated with inductive, capacitance, or resistive components. The phenomena can be DC, alternating current (AC), or transient phenomena.

Examples of different external sources of EOS events include the following:

- **Inductance:** Inductive loads
- **Capacitive:** Cable capacitance charge
- **Resistive:** Ground resistance

Electronic noise in different forms is also a key cause of EOS events. Noise events, both internal and external, can create component failures. Examples of noise events include the following:

- **External switching noise:** Switching noise on antennas
- **External ground plane noise:** Noise on ground plane or current return
- **External EMI:** EMI noise due to poor shielding
- **Internal switching I/O noise:** Sequential switching of digital I/O off-chip driver circuitry
- **Internal switching clock noise:** Switching of timing clocks
- **Internal I/O transients:** Overshoot and undershoot.

1.7.3.1 Characteristic Time Response

ESD event characteristic time response is associated with a specific process of charge accumulation and discharge. Hence, the characteristic time response is definable enough to establish an ESD standard associated with the specific process. Secondly, the time response of ESD events is fast processes. The time constant for ESD events ranges from subnanoseconds to hundreds of nanoseconds.

In contrary, EOS events do not have a characteristic time response [13]. They can have short time response or long (note: today, it is popular to separate the "ESD events" as distinct from "EOS events" which is what will be followed in this text). EOS processes are typically slower and distinguishable from ESD events by having longer characteristic times. The time constant for EOS events ranges from submicroseconds to seconds (Figure 1.5).

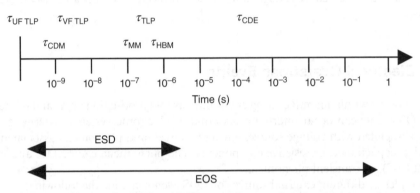

Figure 1.5 EOS and ESD event time constant spectrum.

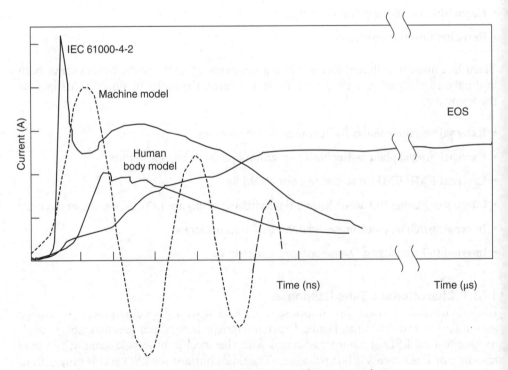

Figure 1.6 EOS and ESD event waveform comparison.

1.7.4 Comparison of EOS versus ESD Waveforms

Figure 1.6 contains both examples of ESD and EOS waveforms. In the plot, ESD waveforms for the HBM, MM, and IEC 61000-4-2 are shown. In comparison, an EOS waveform is highlighted. The key point is that the ESD event waveforms are significantly shorter than EOS events.

1.8 RELIABILITY TECHNOLOGY SCALING AND THE RELIABILITY BATHTUB CURVE

As technologies are scaled, the reliability of semiconductor devices is being affected [13]. This can be observed from the reliability "bathtub" curve. The reliability bathtub curve has three regimes to predict failure rate on a logarithm–logarithm plot of FITs versus time. The FIT rate is the number of fails in 1 billion hours. The first region is known as the infant mortality regime, followed by a second time regime, known as the use or useful life regime, followed by the end-of-life (EOL) regime. The infant mortality is a decreasing linear regime on a log FIT versus log time plot. The second useful life regime is time independent and a low flat FIT rate. As one approaches the EOL regime, reliability "wear-out" begins leading to a linear increase in the FIT rate (Figure 1.7).

As technologies are scaled, premature wear-out is occurring with a continued decrease in the length of the useful life regime. As technologies are scaled from 180 nm to below 65 nm, the length of useful life is decreasing, and wear-out will be a larger issue.

This indicates that the fundamental devices within a semiconductor chip are becoming weaker with technology scaling; it will be important to improve the reliability of components by improving EOS robust circuits through layout, design, topology, and other means to counter the decreasing reliability of semiconductor devices.

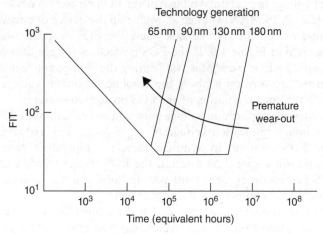

Figure 1.7   Reliability bathtub curve and technology scaling.

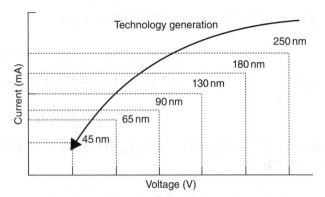

Figure 1.8 Shrinking technology design box.

1.8.1 The Shrinking Reliability Design Box

With technology scaling, the reliability design box is decreasing. Figure 1.8 shows an example of the scaling of the technology reliability design box. With each successive generation, the technologies are getting less robust from a reliability perspective. To compensate for the degradation in the technology device reliability, the solution to provide future EOS robust technology is by providing more EOS robust circuits. EOS robust circuits will be achieved through design layout and circuit topology solutions.

1.8.2 Application Voltage, Trigger Voltage, and Absolute Maximum Voltage

One of the challenges in the development of EOS solution is to develop EOS protection networks whose turn-on voltage is initiated above the application voltage but below the failure voltage of the device or circuit [13]. On the voltage axis, there are an application voltage, a trigger voltage (e.g., clamp voltage) of the EOS protection device, and the absolute maximum (e.g., ABS MAX) voltage allowable on the device or circuit. Hence, there is a desired "window" on the voltage axis where the EOS protection network should operate, as illustrated in Figure 1.9. If the EOS protection voltage turn-on is below the application voltage, the EOS element is "on" during the voltage application range. If the EOS protection voltage turn-on is above the absolute maximum voltage (ABS MAX), then the circuit fails prior to initiation of the EOS protection solution.

When this occurs, the application voltage must address variations in the power supply, V_{DD}, with a maximum application voltage of $V_{DD} + \Delta V_{DD}$. This reduces the triggering window for the EOS solution. In addition, there is temperature variation that also broadens the application space. As a result, the EOS trigger window also is reduced. Hence, the EOS protection element must remain "off" during worst-case voltage and worst-case temperature conditions of the application.

For EOS solutions, another challenge is that there most likely are ESD elements in series with the EOS protection solution, which may also remain off during the application voltage, and must also turn-on below the absolute maximum voltage condition of

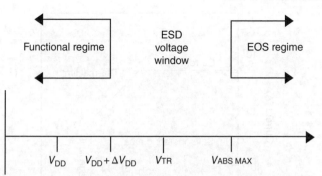

Figure 1.9   Voltage axis highlighting application voltage, EOS protection trigger voltage, and the ABS MAX voltage.

the circuit or device. For power electronics and smart power applications, one of the challenges is to provide a solution for both EOS and ESD protection.

1.9 SAFE OPERATING AREA

Electrical devices, either in integrated electronics or discrete elements, have a region which is regarded as the safe operating area (SOA) in current–voltage (I–V) space [13, 14]. I–V points in the interior of the safe operating I–V space are regarded as states where the device is safe to operate, and I–V points outside of this (SOA) are regarded as a domain where it is regarded as unsafe.

The SOA can be defined from an electrical or thermal perspective. Additionally, one can define a DC SOA or a transient SOA.

Figure 1.10 provides an example of an SOA in the I–V space for a given device.

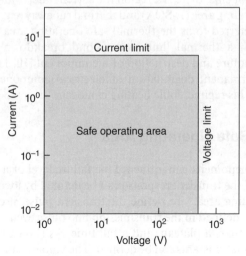

Figure 1.10   Safe operating area.

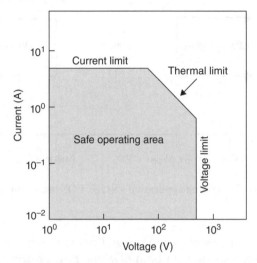

Figure 1.11 SOA with thermal limit.

1.9.1 Electrical Safe Operating Area

Figure 1.11 shows an example of an SOA which includes a thermal limit. A power contour forms a hyperbola on the I-V plot since power equals the product of current times voltage (P=IV). In components and systems, the voltage and current specifications for a rectangle in the *I–V* space. But, due to power limitations, the upper right corner is limited by both power and thermal limits [10, 11, 13, 14].

1.9.2 Thermal Safe Operating Area (T-SOA)

With electronic components, there is a region for current and voltage conditions between the electrical safe operating area (E-SOA) and thermal runaway (e.g., thermal breakdown). This region can be referred to as the thermal safe operating area (T-SOA). Figure 1.12 shows the SOA with a thermal limit and second breakdown limitations. Thermal breakdown leads to failure and destruction of a component [10, 11, 13, 14].

In the T-SOA, permanent degradation of electrical components can occur due to excessive heating. In this regime, Joule heating can occur.

1.9.3 Transient Safe Operating Area

For EOS, transient phenomena can influence the failure level of a device or component. The quantification of the transient response can be defined by identifying and defining a "transient safe operating area." To define the transient pulse, it can be quantified as a trapezoidal pulse as is defined in the transmission line pulse event. The trapezoidal pulse is defined with a rise time, a plateau, and a fall time. As shown in Figure 1.13, voltage states and affiliated time constants can be defined. The voltage can be defined as a plateau voltage, a peak voltage, and a "safe" voltage. For the time constants, corresponding times

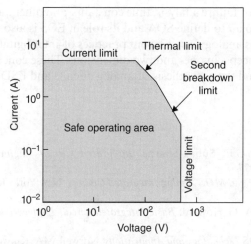

Figure 1.12   SOA with thermal limit and second breakdown limit.

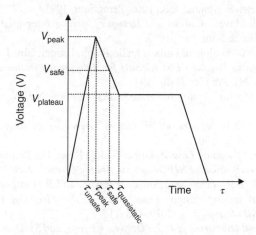

Figure 1.13   Transient safe operating area voltage-time waveform.

can be quantified, such as the unsafe transient time constant (time to the safe voltage), the peak voltage time, the safe voltage time, the quasistatic time (time to the plateau state), as well as the pulse time and the fall time [13].

1.10 CLOSING COMMENTS AND SUMMARY

This chapter opened with the discussion of analog design principles associated with matching and design symmetry. It also discussed EOS and its relationship to other phenomena, such as ESD, EMI, EMC, and latchup. EOS is defined as well in terms of EOC, electrical overpower, and other concepts. In our discussion, there is an emphasis on ESD and EOS events on analog applications. As a result, we have drawn distinctions through the text on difference of failure analysis, time constants, and other means of identification and classification. A plan to define SOA and its role in EOS is also emphasized.

Chapter 2 discusses analog design layout practices of interdigitated design layout and common centroid concepts in one- and two dimensions. These concepts are implemented into ESD networks and the cosynthesis of analog circuits and ESD networks.

REFERENCES

1. A.B. Glasser and G.E. Subak-Sharpe. *Integrated Circuit Engineering*. Reading, MA: Addison-Wesley, 1977.
2. A. Grebene. *Bipolar and MOS Analog Integrated Circuits*. New York: John Wiley & Sons, Inc., 1984.
3. D.J. Hamilton and W.G. Howard. *Basic Integrated Circuit Engineering*. New York: McGraw-Hill, 1975.
4. A. Alvarez. *BiCMOS Technology and Applications*. Norwell, MA: Kluwer Academic Publishers, 1989.
5. R.S. Soin, F. Maloberti, and J. Franca. *Analogue-Digital ASICs, Circuit Techniques, Design Tools, and Applications*. Stevenage, UK: Peter Peregrinus, 1991.
6. P.R. Gray and R.G. Meyer. *Analysis and Design of Analog Integrated Circuits*. 3rd Edition. New York: John Wiley & Sons, Inc., 1993.
7. F. Maloberti. Layout of analog and mixed analog-digital circuits. In: J. Franca and Y. Tsividis (Eds). *Design of Analog-Digital VLSI Circuits for Telecommunication and Signal Processing*. Upper Saddle River, NJ: Prentice Hall, 1994.
8. D.A. Johns and K. Martin. *Analog Integrated Circuit Design*. New York: John Wiley & Sons, Inc., 1997.
9. R. Geiger, P. Allen, and N. Strader. *VLSI: Design Techniques for Analog and Digital Circuits*. New York: McGraw-Hill, 1990.
10. A. Hastings. *The Art of Analog Layout*. Upper Saddle River, NJ: Prentice Hall, 2006.
11. V. Vashchenko and A. Shibkov. *ESD Design for Analog Circuits*. New York: Springer, 2010.
12. H. Kunz, G. Boselli, J. Brodsky, M. Hambardzumyan, and R. Eatmon. An automated ESD verification tool for analog design. *Proceedings of the Electrical Overstress/Electrostatic Discharge (EOS/ESD) Symposium*, 2010; 103–110.
13. S. Voldman. *Electrical Overstress (EOS): Devices, Circuits, and Systems*. Chichester, UK: John Wiley & Sons, Ltd, 2013.
14. S. Voldman. *ESD Basics: From Semiconductor Manufacturing to Product Use*. Chichester, UK: John Wiley & Sons, Ltd, 2012.
15. S. Voldman. *ESD: Physics and Devices*. Chichester, UK: John Wiley & Sons, Ltd, 2004.
16. S. Voldman. *ESD: Circuits and Devices*. Chichester, UK: John Wiley & Sons, Ltd, 2005.
17. S. Voldman. *ESD: RF Circuits and Technology*. Chichester, UK: John Wiley & Sons, Ltd, 2006.
18. S. Voldman. *Latchup*. Chichester, UK: John Wiley & Sons, Ltd, 2007.
19. S. Voldman. *ESD: Circuits and Devices*. Beijing: Publishing House of Electronic Industry (PHEI), 2008.
20. S. Voldman. *ESD: Failure Mechanisms and Models*. Chichester, UK: John Wiley & Sons, Ltd, 2009.
21. M. Mardiquan. *Electrostatic Discharge: Understand, Simulate, and Fix ESD Problems*. Hoboken, NJ: John Wiley & Sons, Inc., 2009.
22. S. Voldman. *ESD: Design and Synthesis*. Chichester, UK: John Wiley & Sons, Ltd, 2011.
23. S. Voldman. Method for evaluating circuit design for ESD electrostatic discharge robustness. U.S. Patent No. 6,526,548, February 25, 2003.

2 Analog Design Layout

2.1 ANALOG DESIGN LAYOUT REVISITED

In analog design, unique design practices are used to improve the functional characteristics of analog circuitry [1–10]. In the electrostatic discharge (ESD) design synthesis of analog circuitry, the ESD design practices must be suitable and consistent with the needs and requirements of analog circuitry [11]. Fortunately, many of the analog design practices are aligned with ESD design practices [12–18]. In the analog design discipline, there are many design techniques to improve tolerance of analog circuits. Analog design techniques include the following:

- Local matching: Placement of elements close together for improved tolerance

- Global matching: Placement in the semiconductor die

- Thermal symmetry: Design symmetry

A key analog circuit design requirement is matching [1]. To avoid semiconductor process variations, matching is optimized by the local placement. Placement within the die location also is an analog concern due to mechanical stress effects. In analog design, there is a concern of the temperature field within the die and the effect of temperature distribution within the die.

Many of the analog design synthesis and practices are also good ESD design practices. The design practices of matching and design symmetry are also suitable practices for ESD design. But there are some design practices where a trade-off exists between the analog tolerance and ESD; this occurs when parasitic devices are formed between the different analog elements within a given circuit or circuit to circuit.

ESD: Analog Circuits and Design, First Edition. Steven H. Voldman.
© 2015 John Wiley & Sons, Ltd. Published 2015 by John Wiley & Sons, Ltd.

In semiconductor development, semiconductor process variation can introduce structural and dimensional nonuniformity. Photolithography and etch tools can introduce these non-uniformities that exist on a local and global design level. These variations can manifest themselves by introducing variations in both active and passive elements. For MOSFET transistors, variation in the MOSFET channel length in single-finger and multifinger MOSFET layouts can lead to nonuniform "turn-on"; this effect can influence both active functional circuits and ESD networks. In bipolar transistors, the linewidth variation can lead to different sizes in emitter structures, leading to nonuniform current distribution in multifinger bipolar transistors. For resistor elements, resistor elements that are utilized for ballasting in multifinger structures can also lead to nonuniform current in the different fingers in the structure.

Design factors that influence the lack of variation are the following semiconductor process and design variables:

- Linewidth

- Line-to-line space

- "Nested-to-isolated" ratio

- Orientation

- Physical spacing between identical circuits

It is an analog circuit design practice and an ESD design synthesis practice to provide a linewidth which is well controlled. For line-to-line space, in an array of lines, the spacing is maintained to provide maximum matching between adjacent lines. For example, in a multifinger MOSFET structure, the spacing between the polysilicon lines is equal to provide the maximum matched characteristics.

Given any array of parallel lines, the characteristics of the "end" or edges of the array can have different characteristics than the other lines. In an array of lines, whereas one edge is adjacent to another line, the other edge is not; this leads to one line-to-line edge space to appear "nested" and the outside line-to-line edge space to appear "semi-infinite" or "isolated." To address the problem of poorly matched edge lines, the following semiconductor process and ESD design solutions are used:

- Process: Cancellation technique of photolithography and etch biases

- Design: Use of dummy edge lines

- Circuit: Use of "gate-driven" circuitry

2.1.1 Analog Design: Local Matching

Analog design local matching is important for segments within a given circuit element or circuit elements within a given circuit [1]. For example, local matching is important in differential pair circuits and current mirrors. Many of the analog design matching practices are also good ESD design practices, which will be discussed.

2.1.2 Analog Design: Global Matching

Analog design global matching is important for circuits that are identical elements but physically spaced around a semiconductor chip [1]. For example, global matching is important in MUX circuits where the inputs are placed at different physical locations within a semiconductor chip. Orientation can also influence the linewidth of identical circuits both locally and globally.

On a macroscopic full-chip scale, there are variations in the photolithography and etching from the top to bottom of a semiconductor chip. In the design of a semiconductor chip, these can be compensated with a preknowledge of the photolithography and etch variations of a technology.

An ESD design practice is to maintain the same x–y orientation of ESD circuits in a semiconductor chip to minimize variation pin to pin. This is not always possible in a peripheral architecture where the ESD element is rotated on the four edges of the semiconductor chip. Note that in this case, the circuit itself (e.g., off-chip driver) may also undergo an orientation effect. It is a good ESD design synthesis practice that addresses the orientation issue with compensation and matching issues for orientation of the ESD elements (in conjunction with the circuit it is protecting).

2.1.3 Symmetry

Symmetry is important to analog design in that it can improve matching characteristics from layout variations and thermal distribution. Analog design global matching is important for circuits that are identical elements but physically spaced around a semiconductor chip.

2.1.4 Layout Design Symmetry

Symmetry is important to analog design in that it can improve matching characteristics from layout variations and thermal distribution [1]. Analog design global matching is important for circuits that are identical elements but physically spaced around a semiconductor chip.

Design symmetry is an ESD design practice to maximize the ESD robustness. The capability of the ESD network to dissipate high-current pulse events is directly related to the network's topology and its design symmetry. The more uniform the current distribution is through the ESD network during a discharge, the better the utilization of the area of the structure, and as a consequence, the greater the robustness of the circuit design. The distribution of current during an ESD event is dependent upon the design symmetry of the ESD network and its components.

To define ESD design symmetry, an axis of symmetry can be defined in the ESD design. Semiconductor design layout is two dimensional, allowing to define an axis of symmetry in the x- and y-direction.

2.1.5 Thermal Symmetry

Thermal symmetry is important in analog design to provide matching of elements in the thermal field [1]. Thermal symmetry is important to avoid mismatch of elements. Since semiconductor element electrical and thermal parameters are temperature dependent (e.g., mobility, electrical conductivity, thermal conductivity), the more uniform the current distribution, the more symmetrical the temperature distribution within the device.

From an ESD perspective, the degree that the design of the ESD network (or structure) on all levels of the integrated circuit departs from a symmetric configuration, the greater is the current localized or nonuniformities in the ESD network. With a symmetrical distribution of the current, the peak power-to-failure per unit area is lowered, producing superior results. Additionally, the more uniform the current distribution, the more uniform the thermal field as well. In integrated circuit design, a key ESD design concept is to maintain a high degree of design symmetry within a structure on all design levels. In both the ESD network and I/O driver circuit, an evaluation of the power distribution of an ESD event within the circuit is an indicator of the robustness of the integrated circuit. Hence, physical layout design symmetry can be used as a heuristic determination of the power distribution within a physical structure.

2.2 COMMON CENTROID DESIGN

Common centroid design practice is used in analog circuitry to provide a high degree of matching [1]. It is commonly used in circuitry where there is a desire to have matching of two components in a circuit. The two devices can be resistors, capacitors, diodes, bipolar junction transistors (BJT), or MOSFET devices.

A typical use of common centroid layout is generally used with differential pair circuitry. It is a matching methodology in which the two transistors of the differential pair circuit are symmetrically laid out about a certain axis. In this fashion, the design style guarantees that both elements see the same semiconductor process variations, providing the best matching characteristics.

2.2.1 Common Centroid Arrays

In the common centroid design method, common centroid arrays are formed to integrate multiple devices in a common area [1]. Arrays whose common centroids align are known as common centroid arrays. Ideally, common centroid arrays eliminate systematic mismatches in the element (Figure 2.1).

2.2.2 One-Axis Common Centroid Design

Common centroid design can be introduced in one dimension or two dimensions [1]. One-dimensional common centroid design is utilized in devices which inherently have a large ratio of length to width. One-dimensional common centroid design is used in the

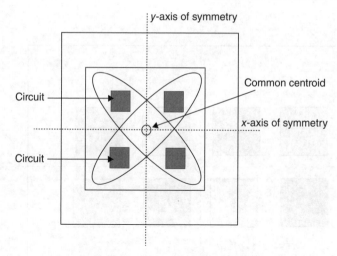

Figure 2.1   Common centroid design.

design of resistors, MOSFET devices, and BJT devices. ESD networks which consist of MOSFETs, BJT, and resistor elements can introduce this design practice.

2.2.3 Two-Axis Common Centroid Design

Common centroid design can be introduced in two dimensions [1]. Figure 2.2 is an example of two-dimensional common centroid design. Two-dimensional common centroid design is introduced in capacitor arrays. Figure 2.3 is an example of a larger two-dimensional centroid array.

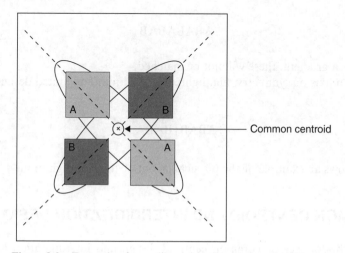

Figure 2.2   Example of two-dimensional common centroid array.

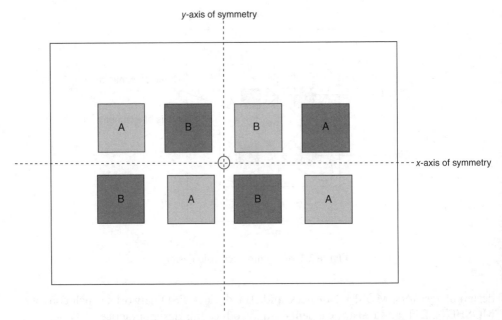

Figure 2.3 Example of a larger two-dimensional common centroid array.

2.3 INTERDIGITATION DESIGN

A second style of layout design for analog is known as interdigitation design layout [1]. In interdigitation layout, devices are in interleaved manner. Capacitors, resistors, diodes, and transistors can be interleaved to provide good matching characteristics and alleviate gradients. For example, given two devices, device A and device B, these can be formed into four unit cells where it is desirable to match device A and device B. This can be represented as

ABABABAB

Given there is a gradient, these will not be matched.

Forming an axis of symmetry, highlighted as "|", an interdigitated design layout can be shown as

ABAB|BABA

Figure 2.4 shows an example of the physical layout of the interdigitation.

2.4 COMMON CENTROID AND INTERDIGITATION DESIGN

The two methodologies of common centroid design and interdigitation are different. First, common centroid design can be in two dimensions. Common centroid design is used when critical matching is required [1, 8, 9].

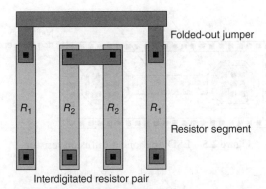

Folded-out jumper

Resistor segment

R_1 R_2 R_2 R_1

Interdigitated resistor pair

Figure 2.4   Example of interdigitation layout.

Interdigitation design layout has an axis of symmetry where two devices are interleaved. Interdigitation design layout eliminates any linear gradient. Common centroid design layout is more precise than interdigitation design but may require more area.

For common centroid design, two axes of symmetry are needed. Assuming device A and device B are divided into 16 unit devices, a common centroid structure, with two axes of symmetry, would be represented as

$$ABAB\,|\,BABA$$
$$\underline{BABA\,|\,ABAB}$$
$$\overline{BABA\,|\,ABAB}$$
$$ABAB\,|\,BABA$$

2.5 PASSIVE ELEMENT DESIGN

Common centroid design and interdigitation design methodologies can be applied to passive elements. Passive elements that utilize these practices can be used in resistors, capacitors, and even inductor elements. The choice of usage of common centroid design versus interdigitation design is a function of the need for critical matching or "good enough" matching characteristics.

2.6 RESISTOR ELEMENT DESIGN

Resistor design is important in analog applications where resistor mismatch is a design issue. Analog design utilizes standard design practices, interdigitated layout design, and common centroid methodologies.

2.6.1 Resistor Element Design: Dogbone Layout

Figure 2.5 is an example of a resistor design element used in analog applications and ESD circuitry [15, 16]. The "dogbone" design style is used in circuitry where a high degree of matching is not required. Figures 2.5 and 2.6 are examples of an n-type dogbone design

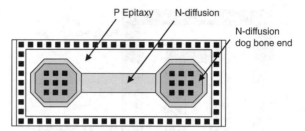

Figure 2.5   ESD dogbone n-diffusion resistor.

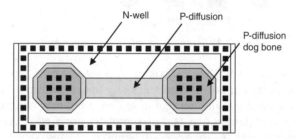

Figure 2.6   ESD dogbone p-diffusion resistor.

and p-type dogbone design, respectively [15, 16]. Dogbone design can be used for ESD input networks where the ends of the dogbone serve as diode elements. In ESD input circuitry that uses a human body model (HBM) dual diode–resistor–charged device model (CDM) dual-diode network, the end of the dogbone can serve as a CDM diode element.

2.6.2 Resistor Design: Analog Interdigitated Layout

Analog applications that require good matching characteristics introduce interdigitated layout design methodologies [1]. Figure 2.4 showed an example of interdigitated design of two resistor elements. In interdigitated designs, connections between resistor segments are electrically connected using "jumpers." Figure 2.7 shows an example of an interdigitated design with "folded-out" jumper elements. Figure 2.8 shows an example of an interdigitated design with "folded-in" jumper elements [1]. Folded-in jumper leads to the interconnect to be placed over the diffusion. For ESD events, this can lead to an increase in the temperature in the region of the interconnect.

2.6.3 Dummy Resistor Layout

Photolithographic and etch variations can lead to width variations of segments in a resistor array. Segments on the edges of arrays can have different widths than segments in the center of an array, due to "nested-to-isolated" width variations. Analog applications that require good matching characteristics can introduce "dummy shapes" on the edges of a resistor segment array. Figure 2.9 is an example of an interdigitated layout design with dummy segments on the edges [1].

In these implementations, the dummy resistor segments can be grounded, biased, or left floating; the electrical connection of the dummy shapes can lead to ESD failure mechanisms between the dummy shape segments and the resistor element.

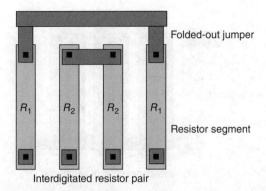

Figure 2.7   Interdigitated resistor layout—folded-out jumper.

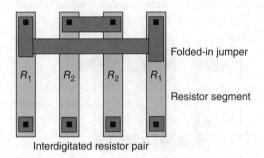

Figure 2.8   Interdigitated resistor layout—folded-in jumper.

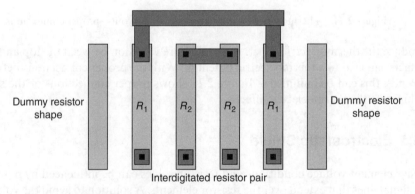

Figure 2.9   Interdigitated resistor layout with dummy resistor shapes.

2.6.4 Thermoelectric Cancellation Layout

The Seebeck effect, also known as the thermoelectric effect, can introduce a thermoelectric potential variation between the two ends of a segment. A thermoelectric potential is equal to the Seebeck coefficient times the temperature difference between the two ends of the resistor segment. The Seebeck coefficient is equal to typically $0.4\,\text{mV/°C}$ where C is the degrees in Celsius [1]. Figure 2.10 shows improper connections of the segments which

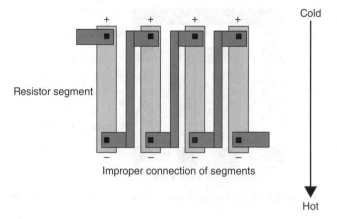

Figure 2.10   Thermoelectric effect in resistors due to resistor layout improper connections.

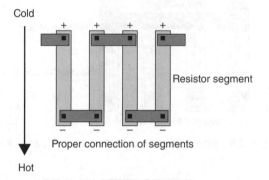

Figure 2.11   Thermoelectric cancellation resistor layout—proper connections.

introduce the thermoelectric effect. A temperature variation between the top and bottom contacts can lead to a thermoelectric potential. With the proper connections between the two ends, this can be minimized. Figure 2.11 shows proper connections of the segments to eliminate the thermoelectric effect.

2.6.5 Electrostatic Shield

Passive element voltage conditions and resistor values can be influenced by power buses and signal lines that extend over the resistor elements. A solution to avoid the voltage bias influence on the passive elements from power bus or signal lines is to introduce "electrostatic shield" or field shield [1]. Figure 2.12 shows an example of an electrostatic shield placed between the overlying power or signal lines and the passive element.

2.6.6 Interdigitated Resistors and ESD Parasitics

Interdigitated resistor elements in differential circuits can lead to parasitic interactions during ESD events. Figure 2.13 provides an example of two interdigitated resistor elements where the resistors are connected to two different signal pins. For the case of a

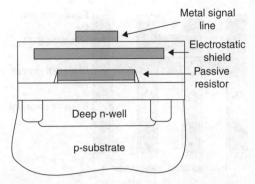

Figure 2.12   Electrostatic shield and resistor layout.

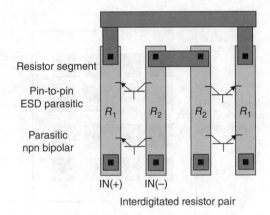

Figure 2.13   Interdigitated resistor layout and ESD parasitic adjacency issue.

differential pair circuit, one resistor is connected to IN(+) and the second resistor is connected to IN(−). For n-type resistors, a parasitic npn bipolar transistor can be formed between the two resistor segments. This parasitic npn transistor can lead to signal pin-to-signal pin ESD failures. At the same time, note that this parasitic npn (with proper design) can be utilized for pin-to-pin ESD protection.

Resistor passive element robustness can be limited by interconnects, such as contacts, vias, and metal lines of the jumper elements. To improve the robustness of the resistor passive and jumper elements, additional contacts or vias can be added to the interdigitated resistor design. ESD failures do occur in resistor elements where a single contact is used. Figure 2.14 provides an example of interdigitated resistor elements with multiple contacts.

2.7 CAPACITOR ELEMENT DESIGN

Capacitors can be designed in a two-dimensional common centroid array where critical matching is needed in analog design [1, 8, 9]. Dummy edge capacitors can be used to avoid edge effects which can influence the capacitance of the elements in the array. Figure 2.15 shows an example of a capacitor array with dummy edge capacitor elements.

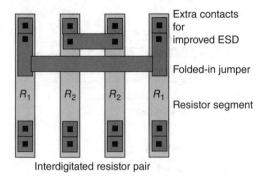

Figure 2.14   ESD robust interdigitated resistor layout with double contacts.

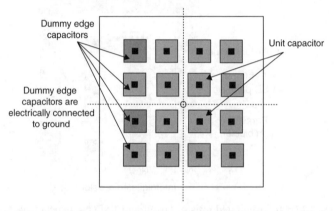

Figure 2.15   Two-dimensional common centroid analog capacitor element with dummy edge capacitors.

2.8 INDUCTOR ELEMENT DESIGN

Inductors are used in analog and radio frequency (RF) applications [17]. On-chip inductors are used on the inputs and output of circuitry. Inductors are used in peripheral circuits for resonant circuits, baluns, and transformers and as AC current blocks (e.g., also known as AC chokes), as well as other circuit applications. High-quality factor inductors are important for analog applications. On-chip inductors in semiconductor technology are constructed from the interconnect technology. The inductors consist of conductive metal films, metal contacts, metal vias, and interlevel dielectrics. The quality factor "Q" can be defined as associated with ratio of the imaginary and real part of the self-admittance term

$$Q = -\frac{\text{Im}\{Y_{11}\}}{\text{Re}\{Y_{11}\}}$$

where, at self-resonance, can be expressed as

$$Q = \frac{\omega}{2} \frac{\partial \varphi}{\partial \omega}\bigg|_{\omega=\omega_{res}}$$

Assuming the quality factor "Q" for an inductor is dependent on the inductor series resistance and the inductance (ignoring capacitance effects), the Q of the inductor with a series resistance is

$$Q = \frac{\omega L}{R}$$

Figure 2.16 is an example of a rectangular inductor design. Figure 2.17 is an example of a symmetric inductor coil pair with a center tap connection. The inductor pair design provides good matching between the two inductor elements.

ESD failures of the inductor elements typically occur at the underpass connection, where the wire cross-sectional area is reduced for the lower-level metal layer. ESD-induced resistance shift can also lead to analog circuit mismatch. In the case that there is a

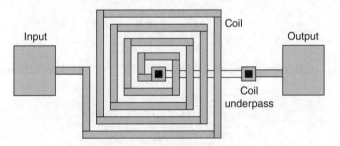

Figure 2.16   Rectangular inductor.

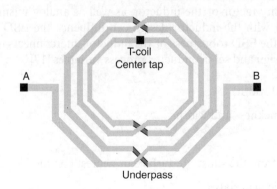

Figure 2.17   Octagonal inductor pair T-coil with center tap.

permanent shift due to ESD stress in the series resistance, post-ESD stress resistance is $R' = R + \Delta R$, and the quality factor "Q" can be expressed as [17]

$$Q' = \frac{\omega L}{R'} = \frac{\omega L}{R + \Delta R}$$

and then the shift in the inductor Q can be expressed as follows:

$$Q' - Q = \frac{\omega L}{R + \Delta R} - \frac{\omega L}{R} = \frac{\omega L}{R}\left[\frac{1}{1 + \dfrac{\Delta R}{R}} - 1\right]$$

Expressing the change in the Q, then

$$\Delta Q = Q' - Q = -\frac{Q}{R + \Delta R}\Delta R$$

where when the $R \gg \Delta R$, then

$$\Delta Q = -\frac{Q}{R}\Delta R$$

For incremental variations, the partial derivative of Q can be taken with respect to resistance, where

$$\frac{\partial}{\partial R}Q = \frac{\partial}{\partial R}\left(\frac{\omega L}{R}\right) = -\frac{Q}{R}$$

From this analysis, changes in the inductor resistance from high-current stress can lead to changes in the quality factor of the inductor, as well as analog mismatch. All physical variables associated with the inductor design can influence the ESD robustness of the inductor structure; the ESD robustness of the inductor interconnects is a function of the following layout design and semiconductor process variables [17]:

- Coil thickness and width
- Underpass film thickness and width
- Via resistance
- Physical distance from the substrate surface or nearest conductive surface
- Interlevel dielectric materials
- Interconnect metal fill material (e.g., aluminum, copper, gold)

- Interconnect cladding material (e.g., refractory metal such as titanium, tantalum, tungsten, etc.)

- Interconnect design (e.g., lift-off, damascene, dual damascene structure)

- Ratio of the volume of the interconnect fill material and volume of the cladding material

- Interconnect design geometry (e.g., square coil, octagonal coil, polygon coil design)

- Interlevel dielectric fill shapes

2.9 DIODE DESIGN

Low-capacitance diode designs that have a high figure of merit of ESD robustness per unit area are desirable for analog applications [13–19]. Circular ESD diode designs are desirable in analog applications because of the following issues:

- Small physical area

- Elimination of isolation and salicide issues

- Elimination of corner effects

- Elimination of wire distribution impact on ESD robustness (e.g., parallel and antiparallel wire distribution issues)

- Current density symmetry

- Integration with octagonal bond pad structures

Circular and octagonal diodes can be placed in small physical areas under bond pads, whether square or octagonal pad structures. Additionally, the small diode structures can be placed in the center of corners of octagonal pads. Isolation issues and corner issues can be eliminated using circular diode structures because of the enclosed nature of the anode or cathode structures. Figure 2.18 shows an example of a diode structure with the p+ anode in the center area; this is separated by an isolation region and an n+ cathode ring structure. As a result, there are no corners in the anode structure which can lead to current concentrations or three-dimensional current distribution effects. Circular diodes have the advantage of avoidance of these geometrical issues. Additionally, due to the physical symmetry, there is a natural design symmetry and no preferred directionality. Additionally, a geometrical advantage of a circular diode and octagonal diode is the radial current flow; the geometrical factor of the $1/r$ current distribution leads to a radially decreasing current density from the center. In linear diode structures, metal distribution can play a role in the nonuniform current distribution; nonuniform current distribution impacts the ESD figure of merit of the ratio of the ESD robustness to capacitance load.

Figure 2.19 is an example of an octagonal design. Octagonal design layout is also preferred over circular layout in some design systems.

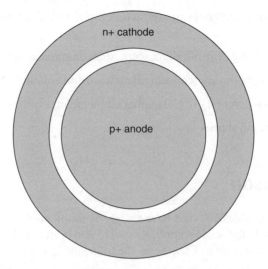

Figure 2.18   Diode layout design—circular design.

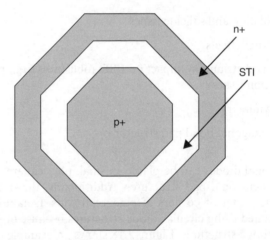

Figure 2.19   Diode layout design—octagonal design.

The disadvantages of an analog circular element are as follows [17]:

- Design width and area limitations.
- Computer-aided design (CAD) issues with nonrectangular shapes.
- CAD automation convergence and process time.
- CAD-parameterized cells for nonrectangular shapes are not always possible preventing variation in the size of the RF ESD element for different circuits or applications.
- Lithography, etch, and polishing issues.
- Limitation on the wire density.

2.10 MOSFET DESIGN

MOSFET device layout and design is critical to both analog design and ESD. For both analog design and ESD robustness, design symmetry and matching is critical. Figure 2.20 is an example of a single-finger MOSFET layout.

It is an analog circuit design practice and an ESD design synthesis practice to provide a linewidth which is well controlled. For line-to-line space, in an array of lines, the spacing is maintained to provide maximum matching between adjacent lines. For example, in a multifinger MOSFET structure, the spacing between the polysilicon lines is equal to provide the maximum matched characteristics. Figure 2.21 shows an example of design symmetry for a MOSFET structure.

Given any array of parallel lines, the characteristics of the "end" or edges of the array can have different characteristics than the other lines. In an array of lines, whereas one edge is adjacent to another line, the other edge is not; this leads to one line-to-line edge space to appear "nested" and the outside line-to-line edge space to appear "semi-infinite"

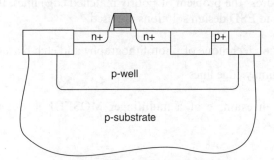

Figure 2.20   Single-finger MOSFET layout.

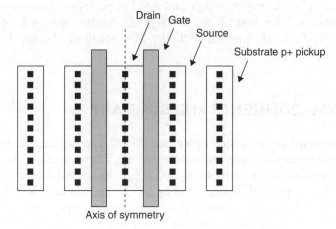

Figure 2.21   Symmetry MOSFET layout.

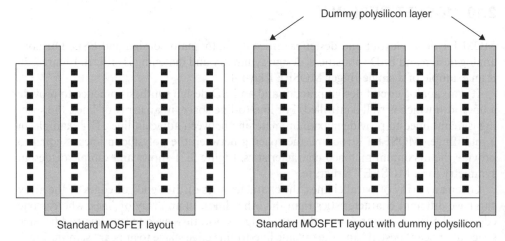

Figure 2.22   Multifinger MOSFET layout with and without dummy lines.

or "isolated." To address the problem of poorly matched edge lines, the following semi-conductor process and ESD design solutions are used:

• Process: Cancellation technique of photolithography and etch biases

• Design: Use of dummy edge lines

Figure 2.22 shows an example of a multifinger MOSFET layout with and without dummy lines.

2.11 BIPOLAR TRANSISTOR DESIGN

In bipolar transistor design, bipolar transistor layout is key for both analog design and ESD [17, 19]. For both analog design and ESD robustness, design symmetry and matching is critical. Analogous to MOSFETs, similar concepts of symmetry, and common centroid, are applied with the placement of the emitter, base, and collector regions.

2.12 CLOSING COMMENTS AND SUMMARY

This chapter discussed analog design layout practices of interdigitated design layout and common centroid concepts in one and two dimensions. These concepts are implemented into ESD networks and the cosynthesis of analog circuits and ESD networks.

In Chapter 3, examples of analog building blocks and circuits that exist in analog designs are provided for readers unfamiliar with analog circuit networks. The analog circuit examples include single-ended receivers, differential receivers, comparators, current mirrors, bandgap regulators, and voltage converters.

REFERENCES

1. A. Hastings. *The Art of Analog Layout*. Upper Saddle River, NJ: Prentice Hall, 2006.
2. A.B. Glasser and G.E. Subak-Sharpe. *Integrated Circuit Engineering*. Reading, MA: Addison-Wesley, 1977.
3. A. Grebene. *Bipolar and MOS Analog Integrated Circuits*. New York: John Wiley & Sons, Inc., 1984.
4. D.J. Hamilton and W.G. Howard. *Basic Integrated Circuit Engineering*. New York: McGraw-Hill, 1975.
5. A. Alvarez. *BiCMOS Technology and Applications*. Norwell, MA: Kluwer Academic Publishers, 1989.
6. R.S. Soin, F. Maloberti, and J. Franca. *Analogue-Digital ASICs, Circuit Techniques, Design Tools, and Applications*. Stevenage, UK: Peter Peregrinus, 1991.
7. P.R. Gray and R.G. Meyer. *Analysis and Design of Analog Integrated Circuits*. 3rd Edition. New York: John Wiley & Sons, Inc., 1993.
8. F. Maloberti. Layout of analog and mixed analog-digital circuits. In: J. Franca and Y. Tsividis (Eds). *Design of Analog-Digital VLSI Circuits for Telecommunication and Signal Processing*. Upper Saddle River, NJ: Prentice Hall, 1994.
9. D.A. Johns and K. Martin. *Analog Integrated Circuit Design*. New York: John Wiley & Sons, Inc., 1997.
10. R. Geiger, P. Allen, and N. Strader. *VLSI: Design Techniques for Analog and Digital Circuits*. New York: McGraw-Hill, 1990.
11. V. Vashchenko and A. Shibkov. *ESD Design for Analog Circuits*. New York: Springer, 2010.
12. H. Kunz, G. Boselli, J. Brodsky, M. Hambardzumyan, and R. Eatmon. An automated ESD verification tool for analog design. *Proceedings of the Electrical Overstress/Electrostatic Discharge (EOS/ESD) Symposium*, 2010; 103–110.
13. S. Voldman. *Electrical Overstress (EOS): Devices, Circuits, and Systems*. Chichester, UK: John Wiley & Sons, Ltd, 2013.
14. S. Voldman. *ESD Basics: From Semiconductor Manufacturing to Product Use*. Chichester, UK: John Wiley & Sons, Ltd, 2012.
15. S. Voldman. *ESD: Physics and Devices*. Chichester, UK: John Wiley & Sons, Ltd, 2004.
16. S. Voldman. *ESD: Circuits and Devices*. Chichester, UK: John Wiley & Sons, Ltd, 2005.
17. S. Voldman. *ESD: RF Circuits and Technology*. Chichester, UK: John Wiley & Sons, Ltd, 2006.
18. S. Voldman. *Latchup*. Chichester, UK: John Wiley & Sons, Ltd, 2007.
19. S. Voldman. *ESD: Failure Mechanisms and Models*. Chichester, UK: John Wiley & Sons, Ltd, 2009.

3 Analog Design Circuits

3.1 ANALOG CIRCUITS

Analog circuits are significantly different from digital circuits from many perspectives. Analog circuits differ from digital networks comprising many basic functional circuit blocks (Figure 3.1). Examples of fundamental analog circuits include the following [1–12]:

- Differential receivers
- Comparators
- Current sources
- Current mirrors
- Voltage regulators
- Bandgap regulators
- Oscillators
- Digital-to-analog converters (DAC)
- Analog-to-digital converters (ADCs)

With these functional circuit blocks, different requirements exist:

- Architecture and domain isolation
- Cross-domain signal line requirements

ESD: Analog Circuits and Design, First Edition. Steven H. Voldman.
© 2015 John Wiley & Sons, Ltd. Published 2015 by John Wiley & Sons, Ltd.

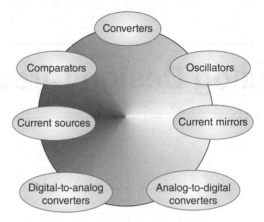

Figure 3.1   Analog design circuits.

- Layout and floor planning restrictions
- Power supply requirements (e.g., low voltage to high voltage)
- Matching requirements
- Symmetry requirements
- Printed circuit board (PCB) placement

3.2 SINGLE-ENDED RECEIVERS

Receiver design and architecture are very important for analog, digital, and radio frequency (RF) applications. Receivers are very sensitive to electrostatic discharge (ESD) events [13–21]. Human body model (HBM) and charged device model (CDM) events can lead to ESD failures. In the following sections, examples of single-ended and differential receivers will be shown.

3.2.1 Single-Ended Receivers

Single-ended receivers are used in both analog and digital receiver designs. Single-ended receivers can be formed with MOSFET devices or bipolar junction transistor (BJT) devices and are used in CMOS, bipolar, and BiCMOS technologies [18]. Figure 3.2 shows a simple single-ended CMOS receiver. CMOS receivers have the MOSFET gate electrically connected to the bond pad. As a result, the MOSFET gate dielectric is vulnerable to external events on the bond pad from electrical overstress (EOS), electrical overvoltage (EOV), or ESD.

Figure 3.3 shows a simple single-ended bipolar receiver. Bipolar receivers have the base electrically connected to the bond pad. As a result, the base–emitter junction is vulnerable to external events on the bond pad from EOS, EOV, or ESD [14].

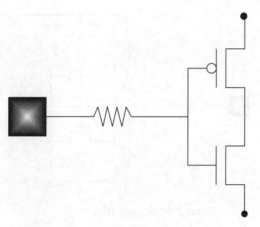

Figure 3.2   CMOS single-ended receiver.

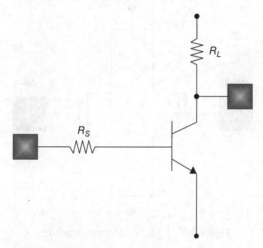

Figure 3.3   Bipolar single-ended receiver.

3.2.2 Schmitt Trigger Receivers

Analog circuitry takes advantage of receiver designs that have hysteresis utilizing feedback networks. Figure 3.4 shows a simple single-ended CMOS Schmitt trigger receiver. In future chapters, the effect of the Schmitt trigger feedback on ESD robustness and the architecture sensitivity will be discussed [17, 18, 20].

3.3 DIFFERENTIAL RECEIVERS

In analog design, differential receivers are used to improve the signal-to-noise ratio in both CMOS and bipolar technologies. In CMOS technology, differential circuits utilize MOSFET transistors in the differential pair, where the signal is connected to the gate electrodes (Figure 3.5) [17, 18, 20].

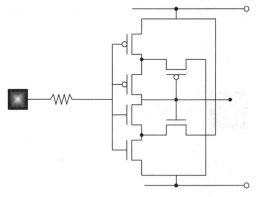

Figure 3.4   Schmitt trigger receiver.

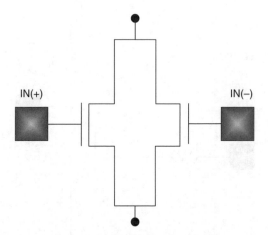

Figure 3.5   CMOS differential receiver.

During ESD events, ESD failures can occur between the differential pair bond pads and the two physical inputs due to signal pin-to-signal pin ESD events. Since there is no electrical current flow through the MOSFET gate structures, the ESD event is associated with either parasitic interactions or the ESD networks.

In bipolar technology, differential receiver networks use a differential pair of identical npn bipolar transistors in a common-emitter mode (Figure 3.6) [18]. For differential bipolar receivers, two input pads are electrically connected to the base contacts of the identical npn transistors, and the two emitters are connected together. Below the emitter connection, an additional circuitry, a current source, or a resistor element is commonly used.

During ESD events, current can flow through the base region to either the base–collector junction or the base–emitter junctions. As a result, interaction between the two sides of the differential network occurs in signal pin-to-signal pin ESD events.

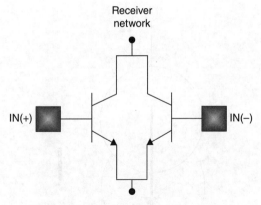

Figure 3.6 Bipolar differential receiver.

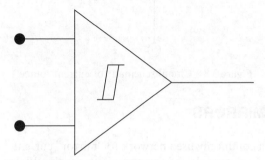

Figure 3.7 Circuit schematic of a comparator.

3.4 COMPARATORS

Comparators are differential circuits whose signal response compares a signal relative to a reference signal level [1–10]. A comparator consists of a specialized high-gain differential amplifier. They are commonly used in devices that measure and digitize analog signals, such as ADCs, as well as relaxation oscillators. Figure 3.7 shows the high-level symbol for a comparator. In analog design, comparators are used to compare a signal to a reference voltage.

3.5 CURRENT SOURCES

Current sources are important in analog design to set up direct current (DC) flow to differential amplifiers, comparators, DAC, and other analog circuits [1–10]. Current mirrors are a form of a current source. Figure 3.8 shows the high-level symbol for a current source.

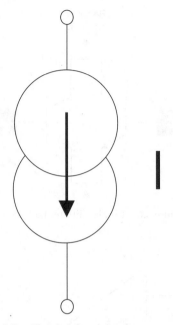

Figure 3.8   Circuit schematic of current source.

3.6 CURRENT MIRRORS

Current mirrors are a commonly used network to "mirror" current from one branch of a network to another. Two examples of current mirror are the Widlar and Wilson current mirrors.

3.6.1 Widlar Current Mirror

A Widlar current source is a modification of the basic two-transistor current mirror that incorporates an emitter degeneration resistor for only the output transistor, enabling the current source to generate low currents using only moderate resistor values. The Widlar circuit may be used with BJT and CMOS MOSFET devices (Figure 3.9) [18, 22]. Figure 3.9 is an example of a Widlar current mirror.

For a bipolar implementation, the Widlar current source includes a resistor and a pair of bipolar transistors having their emitters coupled to opposite ends of the resistor. Additionally, one of the transistors has its collector and base shorted together; the device has a current supply coupled to the shorted collector and base of one of the transistors and to the base of the other transistor. This allows a small predictable difference in voltage to exist between the base and emitter voltages of the transistors which are coupled across the resistor to produce a small output current from the second transistor.

Current mirrors can be contained within a design but also can be electrically connected to a bond pad. In the case of current mirror that is connected to an external pad, the "turn-on" voltage is low, making it difficult to protect from EOS and ESD events.

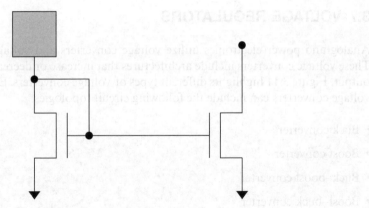

Figure 3.9   Widlar current source.

3.6.2 Wilson Current Mirror

Another implementation of a current mirror used in analog design is a Wilson current mirror [18]. A Wilson current mirror is a three-terminal circuit that accepts an input current at the input terminal and provides a "mirrored" current source or sink output at the output terminal. The mirrored current is a "copy" of the input current. It may be used as a Wilson current source by applying a constant bias current to the input branch. Figure 3.10 shows an example of a Wilson current mirror.

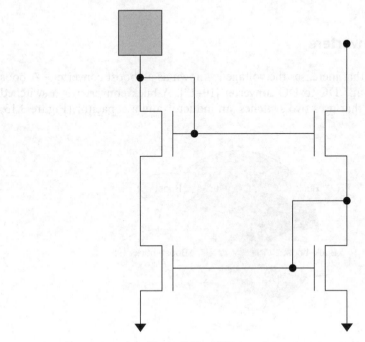

Figure 3.10   Wilson current source.

3.7 VOLTAGE REGULATORS

Analog and power electronics utilize voltage converters and voltage regulators [1–11]. These voltage converters include architectures that increase or decrease the voltage at the output. Figure 3.11 highlights different types of voltage converters. Examples of different voltage converters can include the following circuit topologies:

- Buck converter
- Boost converter
- Buck–boost converter
- Boost–buck converter
- Cuk converters

3.7.1 Buck Converters

A voltage converter that lowers the voltage is known as a "buck converter." A buck converter is a "step-down" DC-to-DC converter [10–12]. A buck converter is a switched-mode power supply that uses two switches, an inductor and a capacitor (Figure 3.12). The two switches can be a transistor and a diode. Protection of the buck converters is important to avoid EOS of the inductor and capacitor at the output, as well as the switch elements.

3.7.2 Boost Converters

A voltage converter that increases the voltage is known as a "boost converter." A boost converter is a "step-up" DC-to-DC converter [10–12]. A buck converter is a switched-mode power supply that uses two switches, an inductor and a capacitor (Figure 3.13).

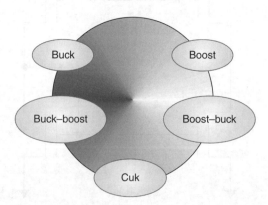

Figure 3.11 Voltage regulators (buck, boost, buck–boost, boost–buck, Cuk).

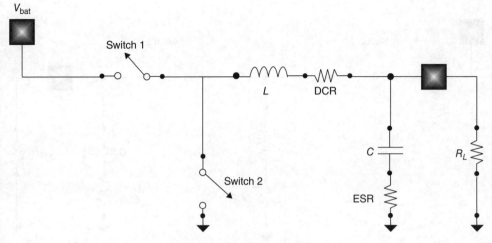

Figure 3.12   Buck converter.

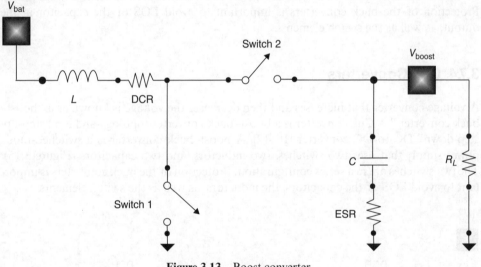

Figure 3.13   Boost converter.

The two switches can be a transistor and a diode. Protection of the buck converters is important to avoid EOS of the capacitor at the output, as well as the switch elements.

3.7.3 Buck–Boost Converters

A voltage converter that decreases and then increases the voltage is known as a "buck–boost converter." A buck–boost converter is a "step-down/step-up" DC-to-DC converter [10–12]. A buck–boost converter is a switched-mode power supply that uses two switches,

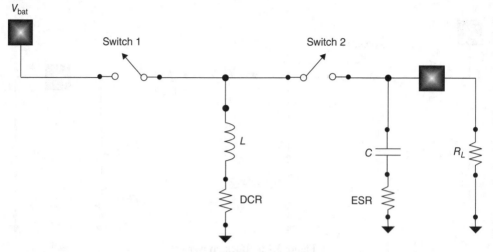

Figure 3.14   Buck–boost converter.

an inductor and a capacitor (Figure 3.14). The two switches are in a series configuration. Protection of the buck converters is important to avoid EOS of the capacitor at the output, as well as the switch elements.

3.7.4 Cuk Converters

A voltage converter that increases and then decreases the voltage is known as a "boost–buck converter." A Cuk converter is a boost–buck converter topology and is a "step-up/step-down" DC-to-DC converter [10–12]. A boost–buck converter is a switched-mode power supply that uses two switches, two inductors, and two capacitors (Figure 3.15). The two switches are in a series configuration. Protection of the buck converters is important to avoid EOS of the capacitors, the inductors, as well as the switch elements.

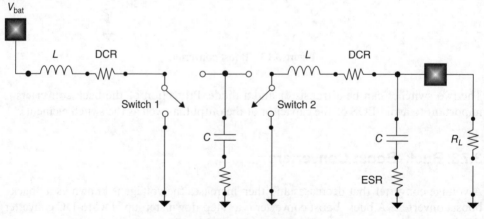

Figure 3.15   Cuk converter.

3.8 VOLTAGE REFERENCE CIRCUITS

Voltage reference circuits are commonly used in analog design [1–12]. These reference circuits must avoid EOS of the internal components to insure proper operation.

3.8.1 Brokaw Bandgap Voltage Reference

A bandgap voltage reference used for application that requires approximately 1.25 V output voltage can utilize the Brokaw bandgap reference network (Figure 3.16) [23]. The circuit maintains an internal voltage source that has a positive temperature coefficient and another internal voltage source that has a negative temperature coefficient. This allows for cancellation of the temperature sensitivity. In the Brokaw bandgap reference,

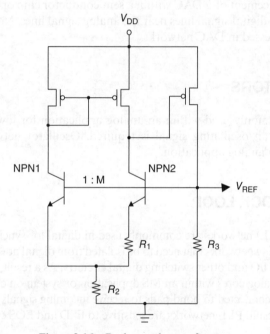

Figure 3.16   Brokaw voltage reference.

the circuit uses negative feedback (e.g., an operational amplifier) to force a constant current through two bipolar transistors with different emitter areas. Bandgap voltage references are important to eliminate the temperature dependence of the voltage reference.

3.9 CONVERTERS

Converters of signal types are important for analog, mixed-signal (MS), and power applications. The two most important converter types are ADCs and DAC [1–12].

3.9.1 Analog-to-Digital Converter

ADCs are used in MS chips to convert analog signals to a digital signal. In an ADC, incoming signals are analog signal lines, and outgoing signals are digital signal lines. Hence, the placement of an ADC within a semiconductor chip or on a PCB is critical to avoid crossing of digital signal lines near the analog signal lines. Noise, ESD, and EOS issues must be addressed in ADC networks.

3.9.2 Digital-to-Analog Converters

DAC are used in MS chips to convert digital signals to an analog signal (Figure 3.17). In a DAC, incoming signals are digital signal lines, and outgoing signals are analog signal lines. Hence, the placement of a DAC within a semiconductor chip or on a PCB is critical to avoid crossing of digital signal lines near the analog signal lines. Noise, ESD, and EOS issues must be addressed in DAC networks.

3.10 OSCILLATORS

Oscillators are commonly used within an analog application for low voltage, or power application, where an oscillating signal is required. Oscillator networks are near the external ports of an analog application.

3.11 PHASE LOCK LOOP

Phase lock loop (PLL) networks are commonly used in digital for synchronization of clock signals. PLL are analog networks that need to be isolated from digital noise. PLL circuits must be away from digital I/O and other switching digital circuitry. As a result, PLL and clocks are isolated from nonanalog cores within an MS application or system on chip (SOC). PLL circuits are electrically connected to bond pads to receive incoming signals to establish the synchronization. As a result, PLL networks are sensitive to ESD and EOS events [17, 18, 20].

3.12 DELAY LOCKED LOOP

Delay locked loop (DLL) networks are commonly used in MS chips and a common functional block for MS chips. The DLL is similar to the PLL which does not have an internal voltage-controlled oscillator. A delay line is used instead of a voltage-controlled oscillator. A DLL circuit may contain delay gates which are connected to a clock. A DLL can be visualized as a negative-delay gate placed in the clock path. Similar to the PLL circuit, this circuit is sensitive to degradation and cannot withstand ESD damage without disruption of the operation. A wide variety of analog applications that utilize feedback exist including analog LED systems (Figure 3.18) to power management unit (PMU) and filter feedback concepts (Figure 3.19).

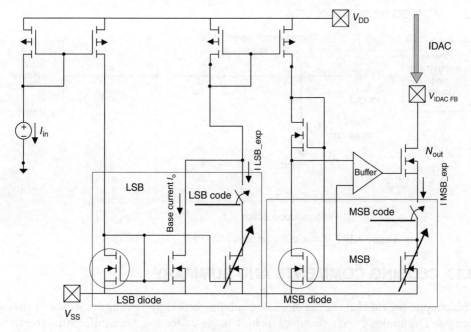

Figure 3.17 Digital to analog converter.

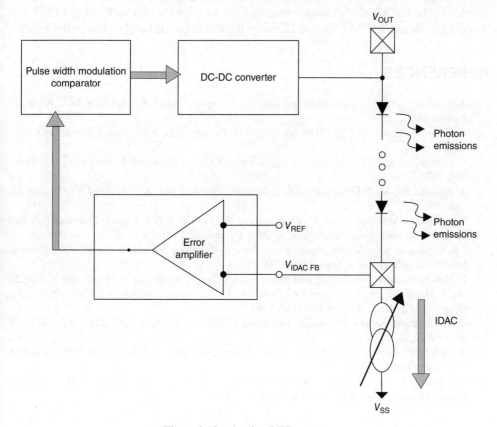

Figure 3.18 Analog LED system.

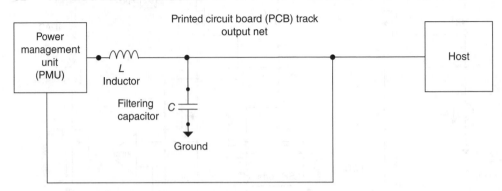

Figure 3.19   Power management unit (PMU) and filter feedback.

3.13 CLOSING COMMENTS AND SUMMARY

This chapter provided examples of analog building blocks and circuits that exist in analog designs. The analog circuit examples include single-ended receivers, differential receivers, comparators, current mirrors, bandgap regulators, and voltage converters.

In Chapter 4, examples of ESD devices used in analog semiconductor chip design are provided. The ESD device examples include both low-voltage and high-voltage ESD networks from diodes, MOSFETs, and LDMOS devices in parallel and series configurations.

REFERENCES

1. A.B. Glasser and G.E. Subak-Sharpe. *Integrated Circuit Engineering*. Reading, MA: Addison-Wesley, 1977.
2. A. Grebene. *Bipolar and MOS Analog Integrated Circuits*. New York: John Wiley & Sons, Inc., 1984.
3. D.J. Hamilton and W.G. Howard. *Basic Integrated Circuit Engineering*. New York: McGraw-Hill, 1975.
4. A. Alvarez. *BiCMOS Technology and Applications*. Norwell, MA: Kluwer Academic Publishers, 1989.
5. R.S. Soin, F. Maloberti, and J. Franca. *Analogue-Digital ASICs, Circuit Techniques, Design Tools, and Applications*. Stevenage, UK: Peter Peregrinus, 1991.
6. P.R. Gray and R.G. Meyer. *Analysis and Design of Analog Integrated Circuits*. 3rd Edition. New York: John Wiley & Sons, Inc., 1993.
7. F. Maloberti. Layout of analog and mixed analog-digital circuits. In: J. Franca and Y. Tsividis (Eds). *Design of Analog-Digital VLSI Circuits for Telecommunication and Signal Processing*. Upper Saddle River, NJ: Prentice Hall, 1994.
8. D.A. Johns and K. Martin. *Analog Integrated Circuit Design*. New York: John Wiley & Sons, Inc., 1997.
9. R. Geiger, P. Allen, and N. Strader. *VLSI: Design Techniques for Analog and Digital Circuits*. New York: McGraw-Hill, 1990.
10. N. Mohan, T. Undeland, and W. Robbins. *Power Electronics: Converters, Applications, and Design*. Hoboken, NJ: John Wiley & Sons, Inc., 2003.

11. I. Batarseh. *Power Electronic Circuits*. Hoboken, NJ: John Wiley & Sons, Inc., 2004.
12. A. Hastings. *The Art of Analog Layout*. Upper Saddle River, NJ: Prentice Hall, 2006.
13. V. Vashchenko and A. Shibkov. *ESD Design for Analog Circuits*. New York: Springer, 2010.
14. S. Voldman. *Electrical Overstress (EOS): Devices, Circuits, and Systems*. Chichester, UK: John Wiley & Sons, Ltd, 2013.
15. S. Voldman. *ESD Basics: From Semiconductor Manufacturing to Product Use*. Chichester, UK: John Wiley & Sons, Ltd, 2012.
16. S. Voldman. *ESD: Physics and Devices*. Chichester, UK: John Wiley & Sons, Ltd, 2004.
17. S. Voldman. *ESD: Circuits and Devices*. Chichester, UK: John Wiley & Sons, Ltd, 2005.
18. S. Voldman. *ESD: RF Circuits and Technology*. Chichester, UK: John Wiley & Sons, Ltd, 2006.
19. S. Voldman. *ESD: Circuits and Devices*. Beijing: Publishing House of Electronic Industry (PHEI), 2008.
20. S. Voldman. *ESD: Failure Mechanisms and Models*. Chichester, UK: John Wiley & Sons, Ltd, 2009.
21. S. Voldman. *ESD: Design and Synthesis*. Chichester, UK: John Wiley & Sons, Ltd, 2011.
22. R.J. Widlar. New developments in IC voltage regulators. *IEEE Journal of Solid State Circuits*, **SC-6**, February 1971; 2–7.
23. P. Brokaw. A simple three terminal IC bandgap reference. *IEEE Journal of Solid State Circuits*, **SC-9** (6), December 1974; 388–393.

4 Analog ESD Circuits

4.1 ANALOG ESD DEVICES AND CIRCUITS

ESD protection for analog design utilizes a wide variety of ESD networks due to a broad spectrum of application spaces. In this chapter, low-voltage to high-voltage ESD devices will be discussed [1–13].

4.2 ESD DIODES

ESD diodes are used for analog design for low-voltage applications from receivers for single-ended and differential pair receivers [2–4]. Many analog applications cannot utilize these devices due to voltage tolerance issues.

4.2.1 Dual Diode and Series Diodes

A common on-chip protection network used in complementary metal-oxide semiconductor (CMOS) technology is the dual-diode (DD) network. With the introduction of the n-channel and p-channel transistor on the same wafer, a protection network using the p- and n-diffusions for diode elements is possible. Figure 4.1 contains the circuit schematic of the DD ESD network [1–4]. This network can be responsive to both electrical overstress (EOS) and ESD events [7, 8]. This network is bidirectional and symmetric turn-on.

ESD: Analog Circuits and Design, First Edition. Steven H. Voldman.
© 2015 John Wiley & Sons, Ltd. Published 2015 by John Wiley & Sons, Ltd.

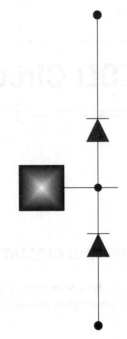

Figure 4.1 DD ESD network.

4.2.2 Dual Diode–Resistor

A common on-chip protection network used in CMOS technology is the DD network with a series resistor. Whereas high-speed digital and radio frequency (RF) applications minimize the series resistance on receivers, analog applications can have resistors with significant resistance magnitude (Figure 4.2) [1–3].

4.2.3 Dual Diode–Resistor–Dual Diode

In CMOS technology, circuits are customized to address specific ESD events. Some of these circuits will also function to address EOS event; but others are not designed for EOS events [1–11]. Figure 4.3 is an example of a circuit to address charged device model (CDM) events. These circuits are used to establish an alternate current path so that the CDM current does not flow through small structures and gated structures. CDM circuitry diverts current from charge that was stored on the V_{DD} power rail or the V_{ss} substrate. These CDM circuits are typically second-stage elements placed close to the failing circuit element. These circuits are low capacitance and small and may not assist for EOS events.

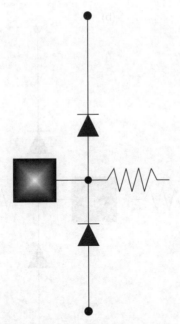

Figure 4.2 DD–resistor ESD network.

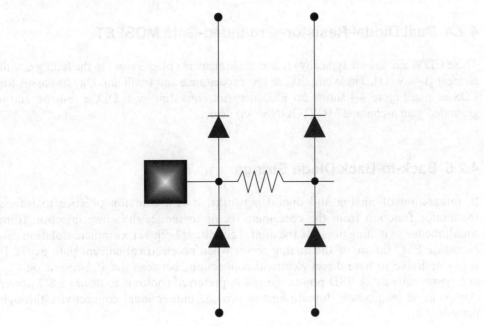

Figure 4.3 DD–resistor–DD ESD network.

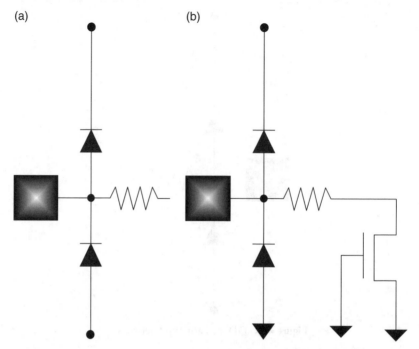

Figure 4.4   (a) DD–resistor ESD network (b) DD–resistor-GGNMOS ESD network.

4.2.4 Dual Diode–Resistor–Grounded-Gate MOSFET

These CDM circuits are typically second-stage elements placed close to the failing circuit element [1–4, 9–11]. These circuits are low capacitance and small and may not assist for EOS events. Figure 4.4 shows an ESD network consisting of a DD, a resistor, and a grounded-gate n-channel MOS (GGNMOS) device.

4.2.5 Back-to-Back Diode Strings

In integration of analog and digital networks, it is a common practice to isolate the analog function from the core logic sector to reduce the noise injection from simultaneous switching noise of the digital circuitry [2–9]. Yet, complete isolation can introduce ESD failure of the analog sector when no electrical current path exists. It is not desirable to have direct electrical connections between the V_{DD}-to-analog V_{DD} and power rails using ESD power clamps. A preferred choice is to utilize ESD power clamps in its own power domain and to provide bidirectional connectivity through the substrate.

In digital circuit applications, the number of diode elements is chosen to allow voltage differential between the ground rails. The placement of ESD networks between ground rails influences both ESD and noise.

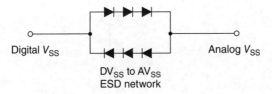

Figure 4.5   Back-to-back symmetric diode string ESD network.

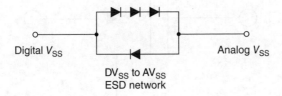

Figure 4.6   Back-to-back asymmetric diode string ESD network.

ESD design practices in the integration of analog and digital chip sectors are as follows:

- Utilize bidirectional element ESD networks between the ground rails of the analog and digital sectors.
- Choose the number of diodes in series based on both the necessary voltage differential and the capacitive coupling requirements.

Typical ESD protection can consist of the following ESD solutions:

- Symmetric back-to-back diode strings
- Asymmetric back-to-back diode strings

4.2.5.1 Back-to-Back Symmetric Diode String
Symmetric back-to-back diode strings are needed in designs that need design symmetric voltage margins or differential circuitry [2–9]. Figure 4.5 shows a back-to-back diode string between the two ground connections between analog and digital core domains.

4.2.5.2 Back-to-Back Asymmetric Diode String
Asymmetric back-to-back diode strings can be used in designs that do not need design symmetric voltage margins Figure 4.6 shows an asymmetric back-to-back diode string between the two ground connections between an analog core and digital core.

4.3 ESD MOSFET CIRCUITS

CMOS technology consists of two active elements—the n-channel metal-oxide semi-conductor field effect transistor (MOSFET) and the p-channel MOSFET. Figure 4.7 shows an example of an n-channel MOSFET structure. An n-channel MOSFET

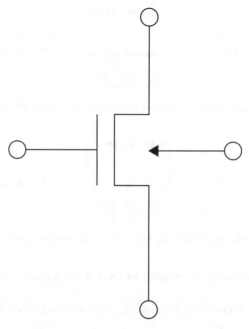

Figure 4.7 CMOS n-channel MOSFET.

consists of a source, drain, gate, and substrate. The source and drain are self-aligned to the MOSFET gate structure. In the n-channel MOSFET, the source and drain are n-doped. The n-channel MOSFET is placed on a p-substrate. The placement of an n-channel MOSFET on a substrate introduces a parasitic p–n diode between the substrate and the MOSFET source and drain; these parasitic diodes can become forward active during ESD and EOS events [1–11].

4.3.1 Grounded-Gate MOSFET

A second common on-chip protection network used in CMOS technology is the grounded-gate n-channel MOSFET network; this is referred to as the GGNMOS protection circuit [2, 4, 7–11, 13]. In this protection network, the MOSFET gate is grounded to keep it off during positive voltage excursions. Figure 4.8 contains the circuit schematic of the GGNMOS network. This network can be responsive to both EOS and ESD events. The network turn-on occurs for positive voltage excursions above the MOSFET-gated breakdown voltage. For negative excursions, the MOSFET n-channel diffusion forward biases to the substrate. This protection is fundamentally bidirectional and asymmetric.

MOSFET devices undergo electrocurrent constriction at high currents. Electrocurrent constriction leads to lower high-current robustness of a MOSFET device. Ballasting techniques are introduced into MOSFET structures to provide uniform current distribution within a MOSFET "finger" as well as MOSFET current distribution finger to finger in a multifinger MOSFET structure. Figure 4.9 shows an example of a silicide block MOSFET layout that will provide a more robust MOSFET device.

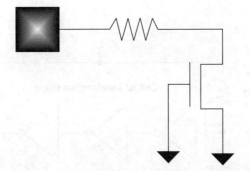

Figure 4.8   Grounded-gate NMOS (GGNMOS).

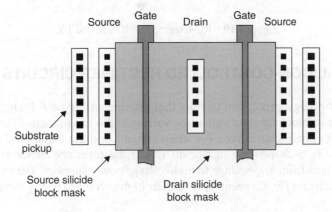

Figure 4.9   Silicide block CMOS MOSFET.

4.3.2 RC-Triggered MOSFET

For analog applications, large MOSFETs are used on signal pins and between power rails. For low-voltage CMOS (LVCMOS) applications, a commonly used ESD network between a power rail and ground rail is the RC-triggered MOSFET circuit [2–11, 13]. In analog power application, this can be used on output nodes for voltage converters and other analog circuit applications.

MOSFETs undergo electrocurrent constriction at high currents. Electrocurrent constriction leads to lower high-current robustness of a MOSFET device; in multifinger structures, nonuniform current distribution leads to the lack of scaling of the MOSFET with MOSFET width in a multifinger MOSFET. Without ballasting techniques, MOSFET current distribution finger to finger in a multifinger MOSFET structure does not always occur. By "gate-driving" the large MOSFET structure, all of the MOSFET fingers will turn on, leading to scaling of the MOSFET operation with MOSFET width.

Using a frequency trigger that responds to the HBM pulse leads to turn-on of the structure from an HBM ESD event, but does not turn on during normal chip operation (Figure 4.10).

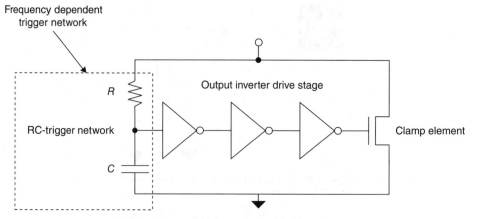

Figure 4.10 RC-triggered CMOS MOSFET.

4.4 ESD SILICON-CONTROLLED RECTIFIER CIRCUITS

In CMOS technology, protection circuits that provide an S-type I–V characteristic can provide a voltage blocking state and a low-voltage/high-current state. Silicon-controlled rectifiers (SCR) provide an S-type I–V characteristic and are commonly used for ESD protection [2–13]. SCR provide significant ESD robustness and hence are desirable in many applications from low-voltage to high-voltage power supplies. They are suitable for electrical overcurrent (EOC) events due to their high current-carrying capabilities.

4.4.1 Unidirectional SCR

SCR are of different classifications and some are desirable for one polarity. Figure 4.11 is an example of a unidirectional SCR that provides an S-type I–V characteristic with a voltage blocking state and a low-voltage/high-current state [2–13].

4.4.2 Bidirectional SCR

Some SCR require bidirectionality due to a symmetric signal swing. Figure 4.12 is an example of a bidirectional SCR that provides an S-type I–V characteristic with a voltage blocking state and a low-voltage/high-current state in both positive and negative polarities [2–13].

4.4.3 Medium-Level Silicon-Controlled Rectifier

In CMOS technology, protection circuits that provide an S-type I–V characteristic can provide a voltage blocking state and a low-voltage/high-current state. SCR provide an S-type I–V characteristic and are commonly used for ESD protection [11]. Figure 4.13

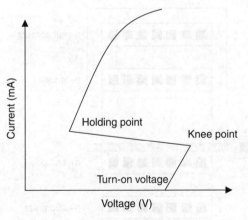

Figure 4.11   Unidirectional SCR network.

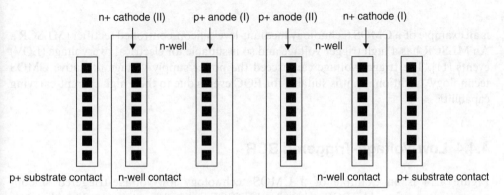

Figure 4.12   Bidirectional SCR network.

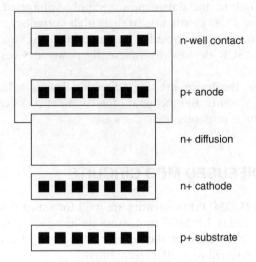

Figure 4.13   MLSCR network.

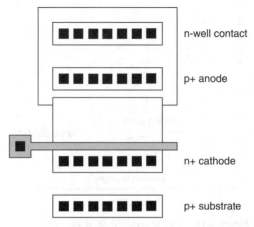

Figure 4.14 LVTSCR network.

is an example of a CMOS technology medium-level silicon-controlled rectifier (MLSCR). An MLSCR has a high trigger voltage and so is suitable for electrical overvoltage (EOV) events [11]. The trigger voltage can exceed the power supply voltage of native CMOS technology. Additionally, it is suitable for EOC events due to their high current-carrying capabilities.

4.4.4 Low-Voltage-Triggered SCR

Figure 4.14 is an example of a CMOS technology low-voltage-triggered silicon-controlled rectifier (LVTSCR) [11]. An LVTSCR has a low trigger voltage by integrating a MOSFET into the pnpn circuit. The advantage of the LVTSCR is that it has a trigger condition closer to the native power supply voltage of the component. LVTSCR are suitable for EOC events due to their high current-carrying capabilities. As a result, these structures are suitable for improving the ESD and EOS protection levels of a component. SCR devices are inherently power devices with good EOS robustness [8].

In the prior discussion, the devices are single direction devices or have an asymmetric response to ESD or EOS events. Bidirectional protection networks can be created that provide a symmetric voltage response.

4.5 LATERALLY DIFFUSED MOS CIRCUITS

Laterally diffused MOS (LDMOS) transistors are used for smart power technology to support higher voltages. Today, LDMOS transistors are being integrated with LVCMOS and bipolar transistors. Integration of the bipolar transistors, LVCMOS, and DMOS transistors is commonly referred to as BCD technology.

4.5.1 LOCOS-Defined LDMOS

Figures 4.15 and 4.16 show examples of the medium-voltage LDMOS (MV-LDMOS) transistor and high-voltage LDMOS (HV-LDMOS) transistors implemented in a LOCOS-isolation technology [6–8, 11, 13]. LOCOS isolation was used in base CMOS technology from 2.0 to 0.8 μm generations to define the MOSFET source and drain regions. In both transistors, the drain region is extended to reduce the surface electric field. This is also referred to as RESURF transistor. In the case of the HV-LDMOS transistor device, the MOSFET gate structure extends over the LOCOS-isolation region. The extension of the gate over the LOCOS isolation lowers the electric field in the MOSFET drain region and decreases gate modulation of the MOSFET drain structure.

An advantage of LOCOS isolation is that the MOSFET junction depth is deeper than the isolation region. When the MOSFET source and drain junctions are deeper than the isolation, current can flow laterally without impediment from the isolation structure. Current crowding can be reduced near the device surface, which leads to a lower temperature internal to the semiconductor chip. A disadvantage of the LOCOS isolation is that the metallurgical junction electric field increases as the MOSFET source and drain junctions are scaled. As the junction is scaled, the radius of the metallurgical junction increases, leading to higher electric field at the junction edge; this leads to lower breakdown voltages. A second disadvantage is the MOSFET channel width control; this leads to MOSFET ΔW variation.

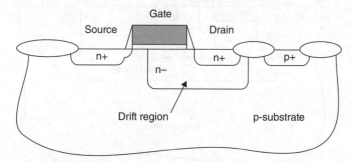

Figure 4.15   LOCOS-defined MV-LDMOS transistor structure.

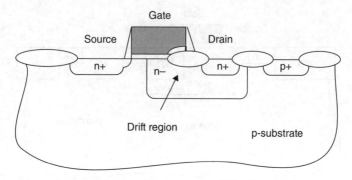

Figure 4.16   LOCOS-defined HV-LDMOS transistor structure.

4.5.2 STI-Defined LDMOS

In base CMOS technology, shallow trench isolation (STI) was integrated to eliminate LOCOS-isolation bird's beak control issues. In STI technology, the MOSFET source and drain junctions are shallower than the isolation depth. From a smart power perspective, STI reduces the lateral current and lateral heat transfer along the device surface. From electrothermal simulation, it is clear that the peak lattice temperature in STI-defined diodes increases (compared to LOCOS-defined diodes) [6–8, 11, 13]. A second concern is the sharp corners introduced in the LDMOS drift region. With technology scaling, STI-defined LDMOS transistors are needed to integrate with the STI-defined LVCMOS technology. Figures 4.17 and 4.18 are a cross section of the STI-defined MV-LDMOS and HV-LDMOS transistors, respectively.

4.5.3 STI-Defined Isolated LDMOS

In LDMOS technology, the application voltages include 125, 50, and 45 V applications. In an LVCMOS STI technology, MOSFET n-type or p-type junction breakdown voltages are typically 12–18 V, and the CMOS n-well-to-substrate breakdown voltages are

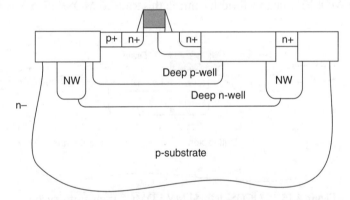

Figure 4.17   STI-defined MV-LDMOS transistor structure.

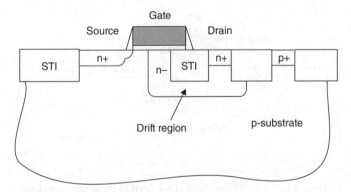

Figure 4.18   STI-defined HV-LDMOS transistor structure.

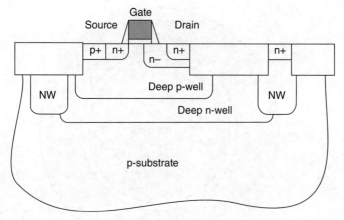

Figure 4.19 STI-defined medium-voltage isolated LDMOS transistor structure.

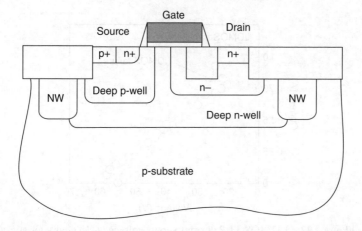

Figure 4.20 STI-defined high-voltage isolated LDMOS transistor structure.

below 45 V. As a result, the n-channel and p-channel low-voltage technology transistors must be isolated from the substrate. In addition, the LDMOS transistors must also have metallurgical junctions which can sustain the higher-voltage applications. LDMOS technology uses deep diffused wells (instead of retrograde implanted well) with high breakdown voltages. These diffused wells can also serve to isolate the LVCMOS transistors from the substrate voltage conditions. Figures 4.19 and 4.20 are a cross section of the isolated STI-defined MV-LDMOS and HV-LDMOS transistors [6–8, 11, 13].

Figure 4.21 is an example layout for a HV-LDMOS. For high-voltage applications, circular design layout provides avoidance of corners and provides symmetry for good current distribution. Figure 4.22 is an example of a transmission line pulse (TLP) I–V characteristic of an LDMOS transistor [6–8, 11, 13].

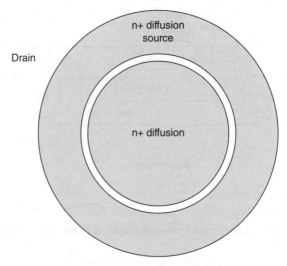

Figure 4.21 LDMOS circular design.

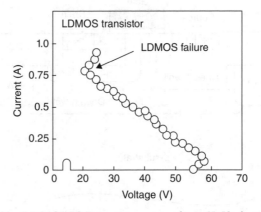

Figure 4.22 LDMOS TLP current versus voltage (I–V) characteristic.

4.6 DEMOS CIRCUITS

High-voltage applications can utilize another type of transistor, known as the drain-extended NMOS (DeNMOS). DeNMOS transistors can also be modified to integrate a SCR within the DeNMOS transistor [7, 8, 11, 13].

4.6.1 DeNMOS

DeNMOS transistor devices introduce a drift region within the drain region to provide a voltage drop within the drain structure. Figure 4.23 shows a cross section of the DeNMOS transistor. The DeNMOS drain region can be modified by extending the distance from the DeNMOS drain contact to gate spacing.

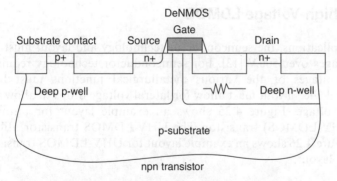

Figure 4.23   DeNMOS transistor cross section.

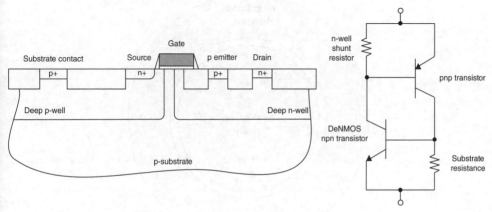

Figure 4.24   DeMOS-SCR transistor cross section.

4.6.2 DeNMOS-SCR

DeNMOS can introduce a p-diffusion region within the drain structure. With the introduction of a p+ diffusion, a pnp can be integrated with the npn transistor to form a DeNMOS-SCR structure. Figure 4.24 shows a cross section of the DeNMOS-SCR transistor. The DeNMOS drain region can be modified by extending the distance from the DeNMOS drain contact to gate spacing. A two-transistor schematic representation is also included in Figure 4.24. In the schematic, the p-well and n-well shunt resistors are also represented. DeNMOS can be used for 40–60 V analog applications.

4.7 ULTRAHIGH-VOLTAGE LDMOS CIRCUITS

In analog design, there are applications that require 600–700 V applied to the semiconductor chip. Ultrahigh-voltage (UHV) technology will require structures where voltages of this magnitude can be applied. In the following section, two types of UHV ESD structures are discussed [13].

4.7.1 Ultrahigh-Voltage LDMOS

For UHV applications, the semiconductor technology and layout must be modified to allow voltages over 500 V [13]. For semiconductor technology requirements, the breakdown voltages of the various metallurgical junctions must be adequate. Secondly, the device layout must allow for lateral voltage drops to allow the distribution of the voltage. Figure 4.25 shows an example layout for ultrahigh-voltage LDMOS (UHV-LDMOS) transistor. The UHV-LDMOS transistor utilizes circular geometry. Figure 4.26 shows an example layout for UHV-LDMOS transistor utilizing a "racetrack" layout.

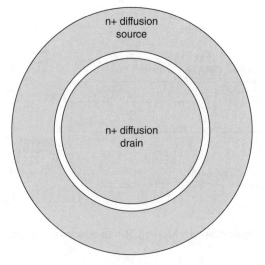

Figure 4.25   UHV-LDMOS layout design: circular design.

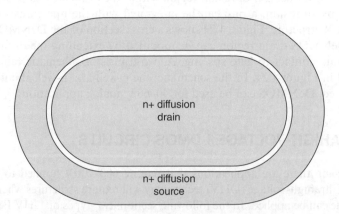

Figure 4.26   UHV-LDMOS layout design: racetrack.

4.7.2 Ultrahigh-Voltage LDMOS SCR

For UHV applications, UHV-LDMOS transistors can be modified to include a p+ diffusion to form a UHV-LDMOS SCR structure [13]. Figure 4.27 shows a cross section of the UHV-LDMOS SCR structure. Figure 4.28 shows an example layout for UHV-LDMOS SCR device utilizing a "racetrack" layout.

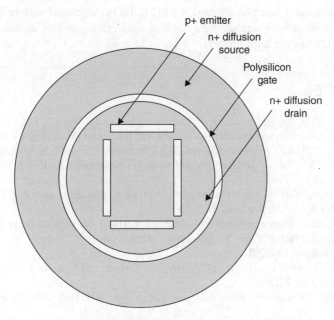

UHV-LDMOS SCR layout

Figure 4.27　UHV-LDMOS SCR layout design: circular layout.

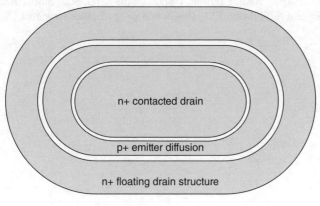

Figure 4.28　UHV-LDMOS SCR layout design: racetrack layout.

4.8 CLOSING COMMENTS AND SUMMARY

This chapter provided examples of ESD devices used in analog semiconductor chip design. The ESD device examples include both low-voltage and high-voltage ESD networks from diodes, MOSFETs, LDMOS, and DeMOS devices in parallel and series configurations. It closed with examples of UHV devices.

In Chapter 5, semiconductor chip architecture in analog and analog-to-digital applications are discussed. Examples of analog ESD failures associated with both circuits and early design architectures are shown. Architectural changes and circuit solutions to address ESD failures are highlighted.

REFERENCES

1. S. Voldman. *ESD: Physics and Devices*. Chichester, UK: John Wiley & Sons, Ltd, 2004.
2. S. Voldman. *ESD: Circuits and Devices*. Chichester, UK: John Wiley & Sons, Ltd, 2005.
3. S. Voldman. *ESD: RF Circuits and Technology*. Chichester, UK: John Wiley & Sons, Ltd, 2006.
4. S. Voldman. *ESD: Failure Mechanisms and Models*. Chichester, UK: John Wiley & Sons, Ltd, 2009.
5. M. Mardiquan. *Electrostatic Discharge: Understand, Simulate, and Fix ESD Problems*. Hoboken, NJ: John Wiley & Sons, Inc., 2009.
6. S. Voldman. *ESD: Design and Synthesis*. Chichester, UK: John Wiley & Sons, Ltd, 2011.
7. S. Voldman. *ESD Basics: From Semiconductor Manufacturing to Product Use*. Chichester, UK: John Wiley & Sons, Ltd, 2012.
8. S. Voldman. *Electrical Overstress (EOS): Devices, Circuits and Systems*. Chichester, UK: John Wiley & Sons, Ltd, 2013.
9. S. Dabral and T.J. Maloney. *Basic ESD and I/O Design*. New York: John Wiley & Sons, Ltd, 1998.
10. A.Z.H. Wang. *On Chip ESD Protection for Integrated Circuits*. New York: Kluwer Publications, 2002.
11. A. Amerasekera and C. Duvvury. *ESD in Silicon Integrated Circuits*. 2nd Edition. Chichester, UK: John Wiley & Sons, Ltd, 2002.
12. A. Hastings. *The Art of Analog Layout*. Upper Saddle River, NJ: Prentice Hall, 2006.
13. V. Vashchenko and A. Shibkov. *ESD Design in Analog Circuits*. New York: Springer, 2010.

5 Analog and ESD Design Synthesis

5.1 EARLY ESD FAILURES IN ANALOG DESIGN

In early development of semiconductor chips, the concepts for implementing analog circuitry on semiconductor chips with other circuitry and providing electrostatic discharge (ESD) protection were not well understood [1–10, 12–18]. Through significant years of ESD design solution, the ESD design synthesis evolved to where it is today [11–18]. In this chapter, examples and case studies of ESD failure in early architecture concepts will be discussed to provide the reader with some historical background and provide insight into the issues that occur in domain separation. The chapter will then discuss the modern-day issues in the present analog architecture.

5.2 MIXED-VOLTAGE INTERFACE: VOLTAGE REGULATOR FAILURES

An early example of voltage regulator ESD failures, in the 1990s, was evident in DRAM applications that required the internal core voltage to be lower than the peripheral circuits [15–17]. In this architecture, the peripheral circuit had a separate power rail and ground distinct from the core power rail and the core ground. Figure 5.1 shows an architecture of the external circuitry and core circuits. Peripheral I/O circuitry would have ESD protection circuits that would be electrically connected to the peripheral I/O power rail [15]. The peripheral I/O power rail was isolated from the core power rail by a voltage regulator. Two different versions of the voltage regulator were implemented.

The voltage regulator can be an n-channel MOSFET or a p-channel MOSFET placed in a drain-to-source configuration between the peripheral I/O and the core power rails. Typically, it is more common to utilize a p-channel MOSFET pass device in the voltage regulator.

ESD: Analog Circuits and Design, First Edition. Steven H. Voldman.
© 2015 John Wiley & Sons, Ltd. Published 2015 by John Wiley & Sons, Ltd.

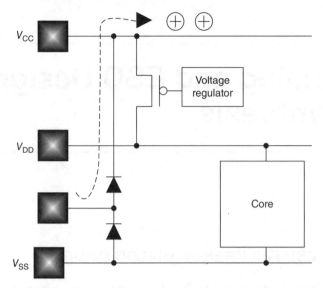

Figure 5.1 ESD architecture with peripheral V_{DD} and core V_{DD} highlighting the analog voltage regulator.

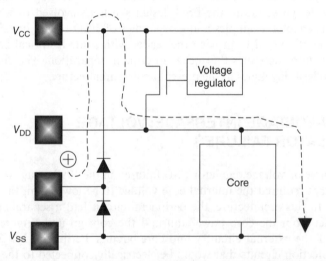

Figure 5.2 ESD current flow through the circuit with an n-channel MOSFET voltage regulator.

In the case of the n-channel MOSFET voltage regulator, ESD failure occurred in the voltage regulator from MOSFET source-to-drain snapback. ESD damage was evident in the MOSFET source, channel, and drain region. Current flowed from the signal pin through the ESD network and through the n-channel MOSFET voltage regulator. The current path of the ESD event is shown in Figure 5.2. The current flowed to the core V_{DD} network [15, 17].

In the case of a p-channel MOSFET voltage regulator, the p-channel MOSFET source and drain are connected between the peripheral I/O and core power rail. The n-well of the p-channel MOSFET is electrically connected to the peripheral I/O. As a result, the p-channel MOSFET source does not forward bias or lead to current flow from the peripheral I/O power rail to the core power rail. In this case, the ESD current flow does not continue to the core power rail, where all the chip capacitance exists. In the case of the p-channel MOSFET low-dropout (LDO) regulator, the ESD current flows through the ESD diode to the peripheral power rail and does not establish a current path to the ground. As a result, the peripheral I/O power rail rises, and the I/O network and the regulator undergo overvoltage, which is a risk to the I/O network and electrical overvoltage (EOV) of the voltage regulator element.

5.2.1 ESD Protection Solution for Voltage Regulator: GGNMOS ESD Bypass between Power Rails

To establish good ESD protection and to avoid electrical overstress (EOS) of the voltage regulator, internal ESD protection can be introduced from the peripheral power rail to the core power rail. The electrical voltage across the voltage regulator must remain under the safe operating area (SOA) voltage limit to avoid damage to the parametric characteristics of the voltage regulator. Figure 5.3 shows an example of an ESD network which is placed in parallel to the voltage regulator, serving as an "ESD bypass network." The ESD bypass network is a grounded-gate NMOS (GGNMOS) device where the MOSFET drain is connected to the peripheral I/O V_{DD} power rail and the MOSFET gate and source are connected to the core V_{DD} power rail. In this fashion, the MOSFET is "off" until the voltage reaches the MOSFET snapback voltage of the ESD bypass network. This allows for the ESD bypass network to be "off" during voltage regulator operation and provides overvoltage protection to the regulator [15, 17].

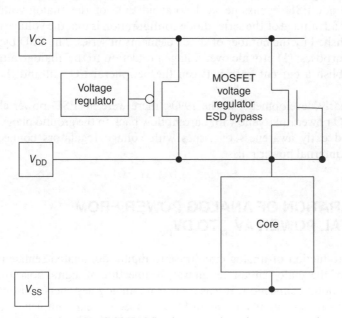

Figure 5.3   ESD GGNMOS voltage regulator bypass network.

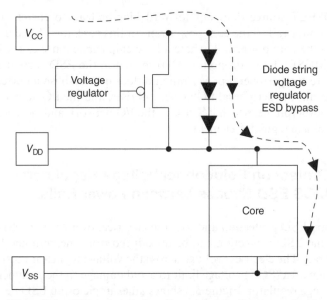

Figure 5.4 ESD series diode voltage regulator bypass network.

5.2.2 ESD Protection Solution for Voltage Regulator: Series Diode String ESD Bypass

An alternate solution for an ESD bypass network which can turn "on" at a lower differential voltage can utilize a series of diode elements. Figure 5.4 shows an example of a series diode ESD bypass network to avoid EOS of the analog voltage regulator element. The advantage of the series diode configuration is that the voltage for "turn-on" can be established by the number of diode elements in series. The ESD bypass network serves two purposes: (1) provide overvoltage protection to the analog voltage regulator and (2) establish a current path between the peripheral I/O rail and the core power rail [15, 17].

In these early developments in the 1990s, there were no ESD power clamps on the peripheral I/O power rail to allow for current flow back to the ground plane. These examples provided early awareness of issues with voltage regulators connected between external and internal power rails.

5.3 SEPARATION OF ANALOG POWER FROM DIGITAL POWER AV$_{DD}$ TO DV$_{DD}$

With the introduction of analog circuitry into digital-dominated semiconductor chips, the concern of the analog circuit design was the injection of digital noise from the power rail, or the ground connection. It was clear from analog designers to isolate the digital noise from the analog circuits. From an ESD perspective, early implementations lead to unique ESD issues within the semiconductor chip [12, 13].

5.4 ESD FAILURE IN PHASE LOCK LOOP (PLL) AND SYSTEM CLOCK

In digital-dominated implementations, there would be a small number of analog circuits within the semiconductor chip. These circuits may include system clocks (e.g., SYS CLOCK) and phase lock loop (PLL) circuitry. These circuits were isolated from the digital power grid by establishing its own local analog V_{DD} (AV_{DD}). The analog power domain may consist of a signal pad and input ESD element, a decoupling capacitor, and the AV_{DD} power rail. In this case, with the AV_{DD} separated from the V_{DD} power rail, the positive polarity ESD event discharged through the diode ESD element to the AV_{DD} power rail. With no current path back to ground, the AV_{DD} power rail charged to a voltage level that leads to EOV of the decoupling capacitor. Electrical failure of the decoupling capacitor leads to an ESD HBM failure level of 1000 V [15–17] (Figure 5.5).

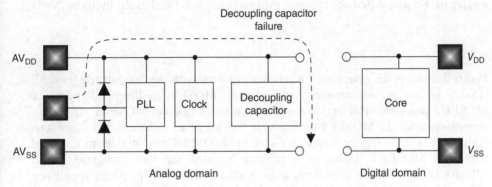

Figure 5.5 Analog domain with system clock, PLL, and decoupling capacitor.

5.5 ESD FAILURE IN CURRENT MIRRORS

Current mirror circuits are commonly used in analog applications. Current mirrors typically are not connected to external signal pads but are in some applications. In a CMOS current mirror network, at least two MOSFET gates are connected to a common node. The current mirror gate is also connected to one of the MOSFET drain regions. Figure 5.6 is an example of a current mirror electrically connected to a signal pad [16].

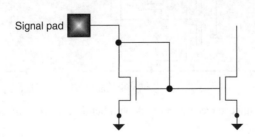

Figure 5.6 CMOS current mirror electrically connected to a signal pad.

ESD protection of the current mirror is challenging since as the voltage rises, the MOSFET current mirror begins to turn on prior to the ESD protection network.

5.6 ESD FAILURE IN SCHMITT TRIGGER RECEIVERS

Receiver networks can introduce feedback elements which provide higher tolerance to noise. Schmitt trigger receiver networks introduce feedback elements to make receivers more tolerant to small input changes. A Schmitt trigger network introduces a hysteresis voltage where the forward and reverse voltage characteristics do not follow the same voltage transfer characteristic path. For the input characteristic, a higher input switching value is needed to switch the circuit when the input value is increasing. This state is referred to as a value of V^+. For the input characteristic, a lower input switching value is needed to switch the circuit when the input value is decreasing. This state is referred to as a value of V^-. The difference between the two states is defined as the hysteresis voltage:

$$V_H = V^+ + V^-$$

Figure 5.7 shows an example of a receiver network with the Schmitt trigger feedback [15–17]. In the receiver network, the p-channel MOSFET pull-up and the n-channel MOSFET pull-down elements are split into a series cascode MOSFET structure. Two elements provide the MOSFET feedback. In the case of an n-channel pull-down stage, an additional Schmitt trigger feedback n-channel MOSFET circuit element is placed; an n-channel MOSFET source is connected between the two n-channel pull-down MOSFETs, and its drain is electrically connected to the local V_{DD} power supply (e.g., or AV_{DD}). The gate of the Schmitt trigger feedback element is connected to the output of the

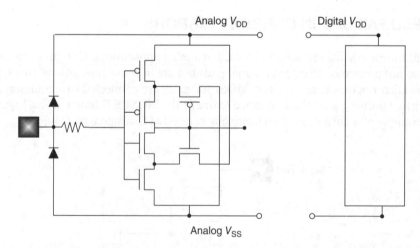

Figure 5.7   Analog symmetric receiver network with Schmitt trigger network feedback in mixed-signal application.

MOSFET receiver output stage. In the case of an p-channel pull-up stage, an additional Schmitt trigger feedback p-channel MOSFET circuit element is placed; a p-channel MOSFET source is connected between the two p-channel pull-up MOSFETs, and its drain is electrically connected to the local V_{SS} power supply (e.g., or analog ground (AV_{SS})). The gate of the p-channel Schmitt trigger feedback element is connected to the output of the MOSFET receiver output stage.

For the receiver network, the two switching conditions can be expressed as [15–17]

$$V^+ = \frac{(V_{DD} + V_{Tn})\sqrt{\dfrac{(W/L)_{n_l}}{(W/L)_{n_f}}}}{1 + \sqrt{\dfrac{(W/L)_{n_l}}{(W/L)_{n_f}}}}$$

where the first width-to-length ratio $(W/L)_{n_l}$ is the lowest receiver n-channel pull-down element and the second width-to-length ratio $(W/L)_{n_f}$ is the n-channel Schmitt trigger feedback element. The reverse trigger voltage is expressed as a function of the p-channel pull-up elements and the p-channel Schmitt trigger feedback element [15–17]:

$$V^- = \frac{(V_{DD} - V_{Tn})\sqrt{\dfrac{(W/L)_{p_l}}{(W/L)_{p_f}}}}{1 + \sqrt{\dfrac{(W/L)_{p_l}}{(W/L)_{p_f}}}}$$

where the first width-to-length ratio $(W/L)_{p_l}$ is the lowest receiver p-channel pull-up element and the second width-to-length ratio $(W/L)_{p_f}$ is the p-channel Schmitt trigger feedback element. A symmetric trigger voltage can be established using this receiver network where

$$V^+ = \frac{1}{2}V_{DD} + \Delta V$$

$$V^- = \frac{1}{2}V_{DD} - \Delta V$$

where the hysteresis is given by

$$V_H = 2(\Delta V)$$

Given that the ratios of the MOSFET receiver stage to the Schmitt trigger feedback for the n-channel and the p-channel elements are the same and assuming a threshold voltage for the n-channel and p-channel is equal in magnitude, it can be expressed as [15–17]

$$\Delta V = \frac{V_{DD}\left(1 - \sqrt{\frac{(W/L)}{(W/L)_f}}\right) + 2V_T\left(1 - \sqrt{\frac{(W/L)}{(W/L)_f}}\right)}{2\left(1 + \sqrt{\frac{(W/L)}{(W/L)_f}}\right)}$$

and $$\sqrt{\frac{(W/L)}{(W/L)_f}} = \frac{V_{DD} - 2(\Delta V)}{V_{DD} + 2(\Delta V) - 2V_T}$$

The MOSFET receiver network is vulnerable from ESD events due to the Schmitt trigger feedback element in the case where the ESD networks are connected to the V_{DD} power rail. Additionally, where this receiver network is placed on an independent power rail, such as analog V_{DD} (e.g., AV_{DD}), ESD failure of Schmitt trigger networks and erratic switching behavior on ESD test systems were first noted. During HBM testing, an ESD diode network can discharge the current to the AV_{DD} power rail. As the ESD current flows to the AV_{DD} power rail, the AV_{DD} power rail voltage increases. As the voltage increases, MOSFET snapback occurs through the Schmitt trigger n-channel MOSFET feedback element and the n-channel MOSFET pull-down element. When the AV_{DD} power rail voltage reaches the voltage condition where the Schmitt trigger and the pull-down MOSFET undergo MOSFET second breakdown, the circuit output failure occurs. From the tester, failure may not be observed since there is no rupture of the MOSFET receiver gate insulator, but the operation of the MOSFET receiver network will not have the same hysteresis character and switching points (Figure 5.8).

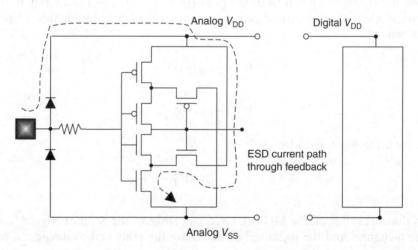

Figure 5.8   Schmitt trigger ESD current path through receiver feedback.

As an ESD practice, feedback elements can lead to early ESD failure of networks. An ESD design practice to prevent failure can be as follows [15–17]:

• **Buffering of feedback elements:** Buffer the feedback elements with series resistance to delay the turn-on.

• **Decoupling of feedback elements:** Decouple feedback elements from the power rails.

• **Alternate current paths to avoid feedback elements:** Establish alternative current paths for the ESD current.

ESD solutions exist which prevent the failure of the MOSFET receiver with the Schmitt trigger feedback elements during ESD events [15–17]:

• Limit the flow of the ESD current to the Schmitt trigger MOSFET along the power bus or series impedance element.

• Provide the MOSFET widths and lengths of the MOSFET Schmitt trigger element and pull-down element to allow for a high MOSFET second breakdown current magnitude.

• Provide ESD power clamps on the power rails which trigger prior to the turn-on of the Schmitt trigger MOSFET feedback and the MOSFET pull-down element.

• Provide a current path to alternate power rails when placed on an independent analog power rail.

In this implementation, the sizing of the MOSFET width-to-length ratio and the relative size of the MOSFET Schmitt trigger elements and the MOSFET pull-up and pull-down elements all influence the hysteresis condition and the triggering point. As a result, cosynthesis of the trigger points, hysteresis condition, and the ESD protection levels are possible by evaluating the size of the MOSFET elements needed in the circuit implementation. Figure 5.9 highlights an ESD improvement in the Schmitt trigger network.

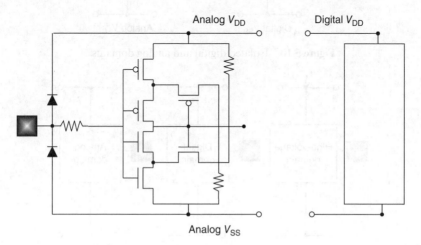

Figure 5.9   Modified Schmitt trigger receiver.

5.7 ISOLATED DIGITAL AND ANALOG DOMAINS

Figure 5.10 shows a chip architecture where the analog and digital domains are isolated. Separation of the digital and analog domains leads to signal pin ESD failures and signal pin-to-signal pin failures [12, 13, 18]. The examples in this chapter are associated with case studies where the analog and digital domains were separated. Figure 5.11 shows an architecture with high-voltage, digital, and analog domain. Whereas the power rails are separated, in a mixed-signal chip, the domains are coupled through the substrate.

5.8 ESD PROTECTION SOLUTION: CONNECTIVITY OF AV_{DD} TO V_{DD}

An ESD solution to the address the analog domain ESD problem is to electrically couple the AV_{DD} to the digital V_{DD} (DV_{DD}) power rail (Figure 5.12). This network can be unidirectional or bidirectional. An example of an inter-rail ESD network can be a single diode element whose anode is connected to the AV_{DD} power rail and the cathode is connected to the V_{DD} power rail. In this fashion, current could flow from the AV_{DD} power rail but rectifying in the reverse direction.

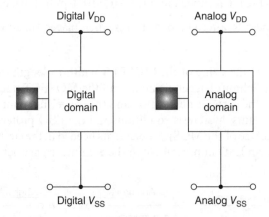

Figure 5.10 Isolated digital and analog domains.

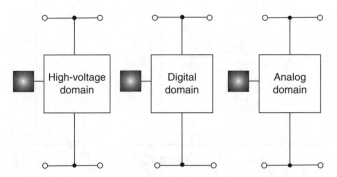

Figure 5.11 Isolated high-voltage, digital, and analog domains.

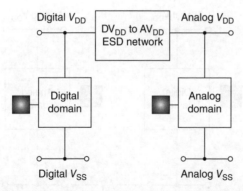

Figure 5.12 Digital to analog with DV_{DD}-to-AV_{DD} ESD network.

This ESD solution was utilized in early semiconductor chip development, but is not commonly used today. The reasons for not using this chip architecture are as follows:

- Power supply voltage differences
- Sequencing
- Power supply noise coupling

5.9 CONNECTIVITY OF AV_{SS} TO DV_{SS}

An ESD solution to address the analog domain ESD problem is to electrically couple the AV_{SS} to the digital ground (DV_{SS}) power rail (Figure 5.13). This network can be unidirectional or bidirectional. An example of an inter-rail ESD network can be a resistor, single diode element, series diodes, diode-configured MOSFETs, or other devices. This ESD solution is more commonly used since there is not an issue with ground potential differentials, or sequencing. Figure 5.14 is an example of interrail coupling using a resistor element.

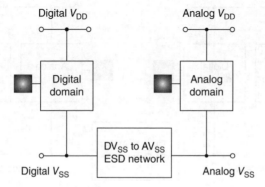

Figure 5.13 Digital to analog with DV_{SS}-to-AV_{SS} ESD network.

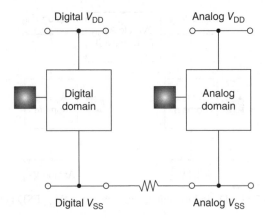

Figure 5.14 DV_{SS}-to-AV_{SS} ESD resistor network.

5.10 DIGITAL AND ANALOG DOMAIN WITH ESD POWER CLAMPS

Today, mixed-signal semiconductor chip architecture contains ESD power clamps in each domain. Figure 5.15 includes an AV_{SS}-to-DV_{SS} power rail ESD network, as well as ESD power clamps between the V_{DD} and V_{SS} of each domain. Figure 5.16 shows the floor plan of a mixed-signal semiconductor chip for the architecture of Figure 5.15. In ESD design synthesis, ESD power clamps are placed in the "pad ring" with the signal bond pads, power bond pads, and power buses [11–13, 18]. In many chip designs, the corners of the semiconductor chip are not utilized. The reasons for not using the corners are as follows:

• Restrictions to place signal pins on the corner.

• Mechanical stress on the chip corners influencing the circuitry.

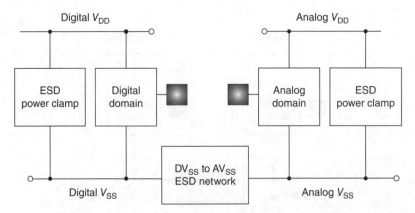

Figure 5.15   Digital and analog domain with ESD power clamps.

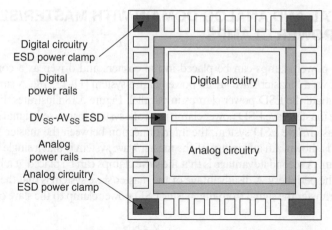

Digital circuitry
ESD power clamp

Digital
power rails

DV_{SS}-AV_{SS} ESD

Analog
power rails

Analog circuitry
ESD power clamp

Digital circuitry

Analog circuitry

Figure 5.16   Floor plan of digital and analog mixed-signal semiconductor chip.

- Photolithography control on the chip corners.
- Placement of identification markings.
- Corners are "white space" regions existing in the semiconductor chips.

It is a common practice in ESD design synthesis to utilize this corner area for placement of the ESD power clamps between the power and ground circuitry.

For mixed-signal designs, two of the corners can used for ESD power clamps for the digital domain (e.g., DV_{DD} and DV_{SS}), and the other two of the corners can be used for the analog power domain ESD power clamps (e.g., AV_{DD} and AV_{SS}) [11–13, 18]. In this architecture, "breaker cells" between the two power domains using ground-to-ground (AV_{SS}-to-DV_{SS}) cells can be utilized. These breaker cells can be placed in the peripheral architecture design. It is typical in these designs that the digital circuits are separated from the analog domains to avoid digital noise from influencing the analog circuitry.

In peripheral I/O design, in very large semiconductor chips or small semiconductor chips that require high ESD robustness, the ESD power clamps are placed at a higher spatial frequency. A natural placement of the ESD power clamps is in the peripheral "standard cell" regions where V_{DD} or V_{SS} power pins are required. In some ASICs, microprocessors, or standard cell foundry methodologies, it is a requirement to place V_{DD} and V_{SS} power pins at a given frequency for a given number of I/O cells. For example, in some methodologies, it is a requirement to place a "power pin" adjacent to every fifth I/O standard cell. Placement of the ESD power clamps within the "power cell" or "power book" allows for the local placement of ESD networks within a given periodicity of every I/O signal pin. Additionally, the placement of the ESD power clamps can be naturally integrated into the design methodology as part of the power pin frequency requirements. In this system, the complete ESD power clamp network is contained throughout the periphery of the semiconductor chip design in a given periodicity.

5.11 DIGITAL AND ANALOG DOMAIN WITH MASTER/SLAVE ESD POWER CLAMPS

Multiple ESD power clamps can be placed in the corners and driven as a common system. Figure 5.17 shows a "master/slave" ESD power clamp system [11–13, 18]. A single trigger network would initiate the ESD power clamps in parallel. Figure 5.18 illustrates the floor plan to incorporate the master/slave ESD power clamp network system in a mixed-signal semiconductor chip. In the master/slave ESD system, the interconnection between the master trigger and the slave networks is shown. An advantage of the master/slave system is that a single trigger initiates the entire system. A second advantage is that the slave clamps can be placed at a high spatial frequency along the periphery. A disadvantage of the master/slave system is that the additional bus is required to transfer the signal from the master ESD power clamp to the slave clamps.

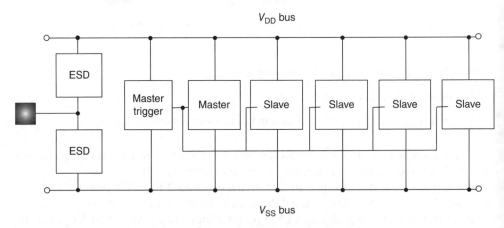

Figure 5.17 Master/slave ESD power clamp system.

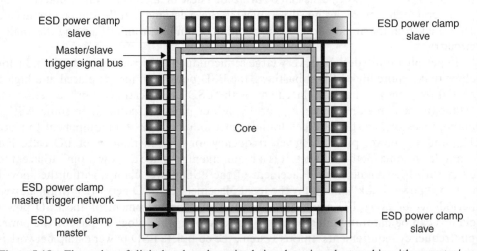

Figure 5.18 Floor plan of digital and analog mixed-signal semiconductor chip with a master/slave ESD power clamp network.

5.12 HIGH-VOLTAGE, DIGITAL, AND ANALOG DOMAIN FLOOR PLAN

Figure 5.19 shows the floor plan of a power technology that contains power devices, as well as digital and analog circuitry. Figure 5.19 highlights the spatial separation of the power, digital, and analog domains. The floor plan also shows the placement of ESD power clamps, power rails, and peripheral I/O [11–13, 18].

Figure 5.20 shows the floor plan of a power technology that contains through-silicon via (TSV) structures between digital and analog circuitry. Figure 5.20 shows signal lines crossing the domain from the digital to analog domain. The TSV structures spatially separate the digital and analog domains. The floor plan also shows the placement of ESD power clamps, power rails, and peripheral I/O [11–13, 18].

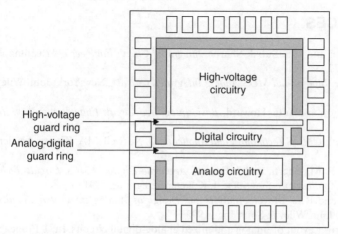

Figure 5.19 Floor plan of high-voltage, digital, and analog mixed-signal semiconductor chip.

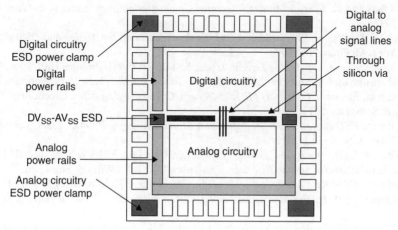

Figure 5.20 Floor plan of digital and analog mixed-signal semiconductor chip with placement of TSV structures.

5.13 CLOSING COMMENTS AND SUMMARY

In this chapter, semiconductor chip architecture in analog and analog-to-digital applications were discussed. Examples of analog ESD failures associated with both circuits and early design architectures are shown. Architectural changes and circuit solutions to address ESD failures are highlighted. Present-day digital-to-analog architectures and floor plans are shown.

Chapter 6 addresses signal line ESD failures in digital and analog domains where they are required to be decoupled due to noise. ESD solutions between the ground connections include coupling using resistors and diode elements, as well as third-party functional blocks. ESD solutions along the signal lines include the ground connections as well as the ESD networks on the signal lines that cross the digital to analog domain.

REFERENCES

1. A.B. Glasser and G.E. Subak-Sharpe. *Integrated Circuit Engineering*. Reading, MA: Addison-Wesley, 1977.
2. A. Grebene. *Bipolar and MOS Analog Integrated Circuits*. New York: John Wiley & Sons, Inc., 1984.
3. D.J. Hamilton and W.G. Howard. *Basic Integrated Circuit Engineering*. New York: McGraw-Hill, 1975.
4. A. Alvarez. *BiCMOS Technology and Applications*. Norwell, MA: Kluwer Academic Publishers, 1989.
5. R.S. Soin, F. Maloberti, and J. Franca. *Analogue-Digital ASICs, Circuit Techniques, Design Tools, and Applications*. Stevenage, UK: Peter Peregrinus, 1991.
6. P.R. Gray and R.G. Meyer. *Analysis and Design of Analog Integrated Circuits*. 3rd Edition. New York: John Wiley & Sons, Inc., 1993.
7. F. Maloberti. Layout of analog and mixed analog-digital circuits. In: J. Franca and Y. Tsividis (Eds). *Design of Analog-Digital VLSI Circuits for Telecommunication and Signal Processing*. Upper Saddle River, NJ: Prentice Hall, 1994.
8. D.A. Johns and K. Martin. *Analog Integrated Circuit Design*. New York: John Wiley & Sons, Inc., 1997.
9. R. Geiger, P. Allen, and N. Strader. *VLSI: Design Techniques for Analog and Digital Circuits*. New York: McGraw-Hill, 1990.
10. A. Hastings. *The Art of Analog Layout*. Upper Saddle River, NJ: Prentice Hall, 2006.
11. V. Vashchenko and A. Shibkov. *ESD Design for Analog Circuits*. New York: Springer, 2010.
12. S. Voldman. *Electrical Overstress (EOS): Devices, Circuits, and Systems*. Chichester, UK: John Wiley & Sons, Ltd, 2013.
13. S. Voldman. *ESD Basics: From Semiconductor Manufacturing to Product Use*. Chichester, UK: John Wiley & Sons, Ltd, 2012.
14. S. Voldman. *ESD: Physics and Devices*. Chichester, UK: John Wiley & Sons, Ltd, 2004.
15. S. Voldman. *ESD: Circuits and Devices*. Chichester, UK: John Wiley & Sons, Ltd, 2005.
16. S. Voldman. *ESD: RF Circuits and Technology*. Chichester, UK: John Wiley & Sons, Ltd, 2006.
17. S. Voldman. *ESD: Failure Mechanisms and Models*. Chichester, UK: John Wiley & Sons, Ltd, 2009.
18. S. Voldman. *ESD: Design and Synthesis*. Chichester, UK: John Wiley & Sons, Ltd, 2011.

6 Analog-to-Digital ESD Design Synthesis

6.1 DIGITAL AND ANALOG

In mixed-signal chips, analog and digital functions are isolated into functional blocks to reduce noise coupling between the digital circuitry and analog functions. The switching noise of the digital circuitry must be isolated to avoid analog functional failures. Figure 6.1 shows an example of a semiconductor chip with separated digital and analog domains. Semiconductor chip architectures today provide the following solutions: (i) separate digital and analog circuit domains, (ii) separate digital and analog V_{DD} power rails, (iii) separate digital V_{SS} (DV$_{SS}$) and analog V_{SS} (AV$_{SS}$) power rails, (iv) an ESD power clamp for each independent domain, and (v) a bidirectional symmetric diode string between DV$_{SS}$ and AV$_{SS}$.

In the ESD design synthesis process, there is a flow of steps and procedures to construct a semiconductor chip. The following design synthesis procedure is an example of an ESD design flow needed for semiconductor chip implementations:

- **I/O, domains, and core floor plan:** Define floor plan of regions of cores, domains, and peripheral I/O circuitry.

- **I/O floor plan:** Define area and placement for I/O circuitry.

- **ESD signal pin floor plan:** Define ESD area and placement.

- **ESD power clamp network floor plan:** Define ESD power clamp area and placement for a given domain.

- **ESD domain-to-domain network floor plan:** Define ESD networks between the different chip domain area and placement for a given domain.

- **ESD signal pin network definition:** Define ESD network for the I/O circuitry.

- **ESD power clamp network definition:** Define ESD power clamp network within a power domain.

ESD: Analog Circuits and Design, First Edition. Steven H. Voldman.
© 2015 John Wiley & Sons, Ltd. Published 2015 by John Wiley & Sons, Ltd.

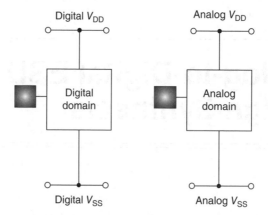

Figure 6.1 Digital and analog domains.

- **Power bus definition and placement:** Define placement, bus width, and resistance requirements for the power bus.

- **Ground bus definition and placement:** Define placement, bus width, and resistance requirements for the ground bus.

- **I/O to ESD guard rings:** Define guard rings between I/O and ESD networks.

- **I/O internal guard rings:** Define guard rings within the I/O circuitry.

- **I/O external guard rings:** Define guard rings between I/O circuitry and adjacent external circuitry.

In this chapter, the focus will be on digital-to-analog interdomain issues.

6.2 INTERDOMAIN SIGNAL LINE ESD FAILURES

In foundry and ASIC environments, chips are segregated into "cores" or subfunctions. Where "cores" are established, the chip is naturally segregated into separate functions, and functional blocks are assembled. In ASIC applications, "voltage islands" are also formed between chip sectors for power management. In this case, many chip sectors' power is shut down, while a given signal path remains "active." In these applications, the power rails are isolated into separate domains, but signal lines pass from one chip subfunction to another; the electrical connectivity between these domains exist only through the common signal paths (and the substrate wafer).

6.2.1 Digital-to-Analog Signal Line Failures

A common interdomain signal line ESD failure mechanism is signal lines between separated subfunctions [1–36]. Figure 6.2 shows the case of two functional blocks, with an inverter where one is the transmitter and the other is the receiver. The ESD networks are

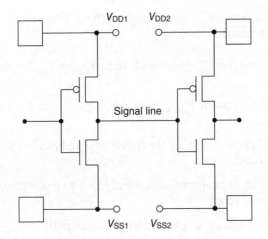

Figure 6.2   ESD interdomain failure.

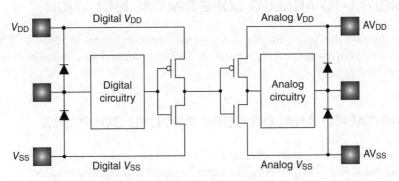

Figure 6.3   Digital-to-analog interdomain issue.

shown in the diagram as well. E. Worley noted that during CDM stress, the receiving inverter network can experience an electrical stress on the MOSFET gate dielectric input [14]. This is due to the resistance of the V_{SS} ground rails, the ESD diode turn-on, and the resistances. Worley noted that the transmitter n-channel MOSFET can transfer current from the charged chip substrate to the transmitter output signal. In that case, current will flow through the signal line between the transmitter and the receiver leading to MOSFET gate dielectric failure in the MOSFET receiver gate.

A common interdomain signal line ESD failure mechanism is signal lines between digital and analog subfunctions [1–14, 34–36]. Digital-to-analog interfaces occur in digital-to-analog converters (DAC) and analog-to-digital converters (ADC). Figure 6.3 shows the case of two functional blocks, with an inverter where one is the transmitter and the other is a receiver. The ESD networks are shown in the diagram as well. As in Figure 6.2, the receiving inverter network can experience an electrical stress on the MOSFET gate dielectric input. Current will flow through the signal line between the transmitter and the receiver leading to MOSFET gate dielectric failure in the MOSFET receiver gate.

In this concept, proposed solutions to address these failure mechanisms, the following solutions were recommended:

- Resistor elements between V_{DD} power rails (e.g., V_{DD} to V_{DD}) as well as V_{SS} power rails (e.g., V_{SS} to V_{SS})

- Back-to-back diodes between V_{DD} power rails (e.g., V_{DD} to V_{DD}) as well as V_{SS} power rails (e.g., V_{SS} to V_{SS})

- Adding CDM ESD circuits between the transmitter and receiver networks providing an "internal" ESD network

- Adding an additional "third-party" functional block to transmit signals between a first and second functional block

- Using additional intervening blocks within the signal path

6.3 DIGITAL-TO-ANALOG CORE SPATIAL ISOLATION

Figure 6.4 shows the floor plan of a mixed-signal semiconductor chip [34–36]. The digital and the analog cores are spatially separated. This minimizes digital noise from impacting analog circuitry.

6.4 DIGITAL-TO-ANALOG CORE GROUND COUPLING

Figure 6.5 illustrates a circuit schematic of an ESD network between the digital and the analog grounds (e.g., DV_{SS} to AV_{SS}) in a mixed-signal semiconductor chip.

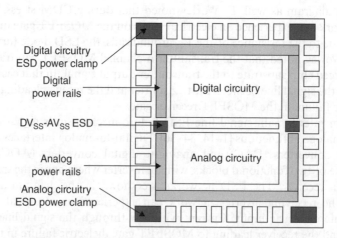

Figure 6.4   Floor plan of digital and analog domains.

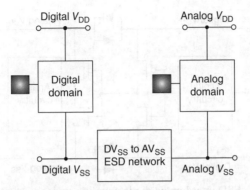

Figure 6.5   Circuit schematic of digital and analog domains with DV_{SS} to AV_{SS}.

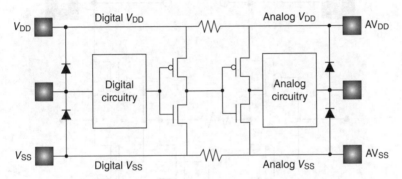

Figure 6.6   Analog-to-digital architecture with domain-to-domain resistor coupling.

6.4.1 Digital-to-Analog Core Resistive Ground Coupling

Figure 6.6 shows an example of an analog-to-digital domain with resistor elements between the power supplies. Resistive coupling allows for the control of the resistance between the two chip subfunctions through design.

6.4.2 Digital-to-Analog Core Diode Ground Coupling

Analog-to-digital subfunctions place antiparallel diode strings between the ground power rails for ESD protection [34–36]. For ESD protection, it is common to place antiparallel diode strings between the electrical grounds to establish ESD current paths between the V_{SS} power rails (e.g., AV_{SS} and DV_{SS}). Figure 6.7 highlights an architecture where the V_{SS}-to-V_{SS} ESD power rail is an antiparallel diode string. In the placement of the antiparallel diode strings, the bus resistance of the V_{SS} and the V_{DD} can play a role in the voltage stress across the signal lines. In the case of signal lines between functional blocks, a voltage differential can be established between the output of a first logic block and the input of the second logic block.

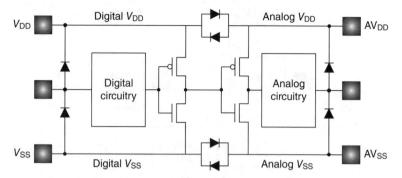

Figure 6.7 ESD interdomain diode-coupling analog-to-digital solution.

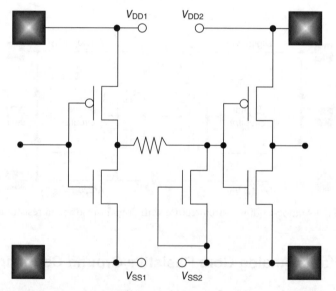

Figure 6.8 ESD interdomain CDM failure using resistor and grounded-gate NMOS (GGNMOS) circuit.

6.5 DOMAIN-TO-DOMAIN SIGNAL LINE ESD NETWORKS

Figure 6.8 shows an example of an interdomain ESD network. For CDM solutions, the second-stage network consists of a resistor and dual-diode element [14]. Other CDM networks can consist of a resistor and MOSFET element.

6.6 DOMAIN-TO-DOMAIN THIRD-PARTY COUPLING NETWORKS

Voltage drops along cross-domain signal lines can be achieved by introduction of additional circuitry along the signal path. A "third-party" functional block to transmit the signals between the transmitter and the receiver can be introduced [14]. This third-party

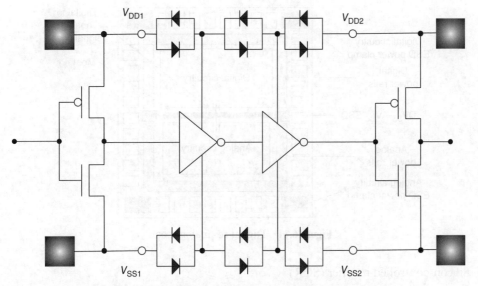

Figure 6.9   ESD interdomain third-party functional block circuit.

block contains two inverters in series, supportive ESD elements, and its own independent power rails (Figure 6.9).

6.7 DOMAIN-TO-DOMAIN CROSS-DOMAIN ESD POWER CLAMP

A means to avoid current flow from the digital power rail (DV_{DD}) and the analog ground rail (AV_{SS}) can be achieved by introduction of a "cross-domain" ESD network [34–36]. Figure 6.10 illustrates a circuit schematic highlighting the cross-domain ESD network. The cross-domain ESD network can be the following:

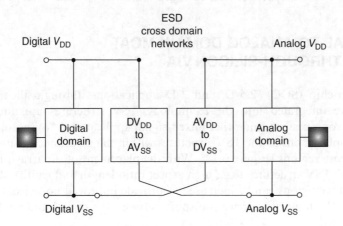

Figure 6.10   ESD cross-domain power clamp.

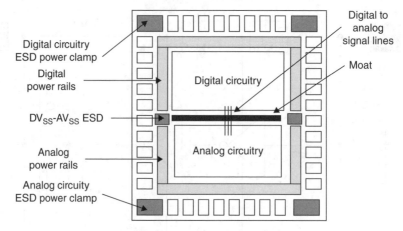

Figure 6.11   Digital-to-analog moat.

- Silicon-controlled rectifier (SCR)

- GGNMOS

- Series diodes

- RC-triggered MOSFET power clamp

6.8 DIGITAL-TO-ANALOG DOMAIN MOAT

In mixed-signal applications, noise isolation and substrate injection are critical for analog circuitry. Noise isolation can be improved by placement of a "moat" that isolated the digital and analog core regions (Figure 6.11) [34–36]. With the placement of a "moat," the resistance between the two regions can be increased to provide improved noise isolation. Additionally, the minority carrier transport between the digital and analog domains can be decreased.

6.9 DIGITAL-TO-ANALOG DOMAIN MOAT WITH THROUGH-SILICON VIA

In system-on-chip (SOC), 2.5-D, and 3-D applications, through-silicon via (TSV) structures are integrated into the technology. TSV structures are integrated into memory, microprocessors, and other mixed-signal applications. Noise isolation can be improved by placement of the TSV structures within the moat that isolated the digital and analog core regions (Figure 6.12). With the placement of an array of TSV structures or "bar" TSV structures (e.g., high aspect ratio length and width), the resistance between the two regions can be increased to provide improved noise isolation [37–45]. Additionally, the minority carrier transport between the digital and analog domains can be decreased.

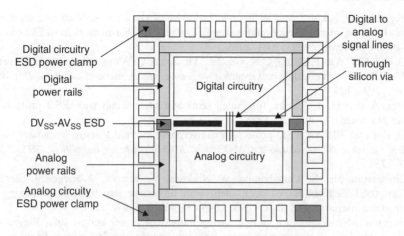

Figure 6.12   Digital-to-analog floor plan with TSV within the moat region.

6.10 DOMAIN-TO-DOMAIN ESD DESIGN RULE CHECK AND VERIFICATION METHODS

ESD checking and verification are used to address finding the cross-domain signal lines to avoid ESD failure [15–33]. These ESD design checking used various methods that address circuit topology connectivity and resistance along the signal lines that cross the digital-to-analog domain. This will be discussed in Chapter 12.

6.11 CLOSING COMMENTS AND SUMMARY

This chapter addressed integration of digital and analog domains where they are required to be decoupled due to noise. Electrical isolation of the two domains leads to ESD failures between the grounds, as well as along signal lines that cross these separated domains. ESD solutions between the ground connections include coupling using resistors and diode elements, as well as third-party functional blocks. ESD solutions along the signal lines include the ground connections as well as the ESD networks on the signal lines that cross the digital-to-analog domain.

Chapter 7 addresses analog and ESD signal pin cosynthesis that introduces usage of interdigitated layout and common centroid concepts to provide ideal matching, low capacitance, and small area in differential signal pins. Using interdigitated designs, the parasitic elements are utilized for signal pin-to-signal pin ESD protection.

REFERENCES

1. C. Johnson, T.J. Maloney, and S. Qawami. Two unusual HBM ESD failure mechanisms on a mature CMOS process. *Proceedings of the Electrical Overstress/Electrostatic Discharge (EOS/ESD) Symposium*, 1993; 225–231.

2. H. Terletzki, W. Nikutta, and W. Reczek. Influence of the series resistance of on-chip power supply buses on internal device failure after ESD stress. *IEEE Transactions on Electron Devices*, **ED-40**(11), November 1993; 2081–2083.

3. M.D. Ker, C.Y. Wu, T. Cheng, C.N. Wu, T.L. Yu, and A.C. Wang. Whole chip ESD protection for VLSI/ULSI with multiple power pins. *Proceedings of the International Reliability Workshop (IRW)*, 1994; 124–128.

4. W. Reczek and H. Terletzski. Integrated semiconductor circuit with ESD protection. U.S. Patent No. 5,426,323, June 20, 1995.

5. M.D. Ker and T.L. Yu. ESD protection to overcome internal gate-oxide damage on digital-analog interface of mixed-mode CMOS IC's. *Microelectronics Reliability*, **36**(11/12), 1996; 1727–1730.

6. I. Chyrsostomides, X. Guggenmos, W. Nikkuta, W. Reczek, J. Rieger, J. Stecker, and H. Terletzski. Semiconductor component with protective structure for protecting against electrostatic discharge. U.S. Patent No. 5,646,434, July 8, 1997.

7. H. Imamura. Semiconductor device having individual power supply lines shared between function blocks for discharging surge without propagation of noise. European Patent Application 97108486.8, March 12, 1997.

8. M.D. Ker, C.Y. Wu, H.H. Chang, and T.S. Wu. Whole chip ESD protection scheme for CMOS mixed-mode IC's in deep submicron CMOS technology. *Proceedings of the Custom Integrated Circuits Conference (CICC)*, May 5–8, 1997; 31–38.

9. H. Nguyen and J.D. Walker. Electrostatic discharge protection system for mixed voltage application specific integrated circuit design. U.S. Patent No. 5,616,943, April 1, 1997.

10. E. Worley, C.T. Nguyen, R.A. Kjar, and M. Tennyson. Method and apparatus for coupling multiple independent on-chip V_{DD} busses to an ESD core clamp. U.S. Patent No. 5,654,862, August 5, 1997.

11. W. Nikkuta and W. Reczek. Integrated semiconductor circuit. U.S. Patent No. 5,821,804, October 13, 1998.

12. P. Minogue. ESD protection scheme. U.S. Patent No. 5,731,940, March 24, 1998.

13. M.D. Ker. ESD protection circuit for mixed mode integrated circuits with separated power pins. U.S. Patent No. 6,075,686, June 13, 2000.

14. E.R. Worley. Distributed gate ESD network architecture for inter-power domain signals. *Proceedings of the Electrical Overstress/Electrostatic Discharge (EOS/ESD) Symposium*, 2004; 238–247.

15. C. Robertson. Calibre PERC: Preventing electrical overstress failures. *EE Times Design, EE Times*, February 9, 2012.

16. TSMC Open Innovation Platform Ecosystem Forum. Improving analog/mixed signal circuit reliability at advanced nodes. *EDA Technical Presentations*, 2011.

17. M. Khazhinsky. ESD electronic design automation checks. *In Compliance Magazine*, August 2012.

18. M. Hogan. Electronic design automation checks, Part II: Implementing ESD EDA checks in commercial tools. *In Compliance Magazine*, September 2012.

19. S. Sinha, H. Swaminathan, G. Kadamati, and C. Duvvury. An automated tool for detecting ESD design errors. *Proceedings of the Electrical Overstress/Electrostatic Discharge (EOS/ESD) Symposium*, 1998; 208–217.

20. M. Baird and R. Ida. Verify ESD: A tool for efficient circuit level ESD simulations of mixed signal ICs. *Proceedings of the Electrical Overstress/Electrostatic Discharge (EOS/ESD) Symposium*, 2000; 465–469.

21. P. Ngan, R. Gramacy, C.K. Wong, D. Oliver, and T. Smedes. Automatic layout based verification of electrostatic discharge paths. *Proceedings of the Electrical Overstress/Electrostatic Discharge (EOS/ESD) Symposium*, 2001; 96–101.

22. R. Zhan, H. Xie, H. Feng, and A. Wang. ESDZapper: A new layout-level verification tool for finding critical discharging path under ESD stress. *Proceedings of the ASPDAC*, 2005; 79–82.

23. H.Y. Liu, C.W. Lin, S.J. Chou, W.T. Tu, C.H. Liu, Y.W. Chang, and S.Y. Kuo. Current path analysis for electrostatic discharge protection. *Proceedings of the ICCAD*, 2006; 510–515.

24. J. Connor, S. Mitra, G. Wiedemeir, A. Wagstaff, R. Gauthier, M. Muhammad, and J. Never. ESD simulation using fully extracted netlist to validate ESD design improvement. *Proceedings of the International ESD Workshop (IEW)*, 2007; 396–407.

25. T. Smedes, N. Trivedi, J. Fleurimont, A.J. Huitsing, P.C. de Jong, W. Scheucher, and J. van Zwol. A DRC-based check tool for ESD layout verification. *Proceedings of the Electrical Overstress/Electrostatic Discharge (EOS/ESD) Symposium*, 2009; 292–300.

26. Z.Y. Lu and D.A. Bell. Hierarchical verification of chip-level ESD design rules. *Proceedings of the Electrical Overstress/Electrostatic Discharge (EOS/ESD) Symposium*, 2010; 97–102.

27. H. Kunz, G. Boselli, J. Brodsky, M. Hambardzumyan, R. Eatmon. An automated ESD verification tool for analog design. *Proceedings of the Electrical Overstress/Electrostatic Discharge (EOS/ESD) Symposium*, 2010; 103–110.

28. M. Okushima, T. Kitayama, S. Kobayashi, T. Kato, and M. Hirata. Cross domain protection analysis and verification using whole chip ESD simulation. *Proceedings of the Electrical Overstress/Electrostatic Discharge (EOS/ESD) Symposium*, 2010; 119–125.

29. M. Muhammad, R. Gauthier, J. Li, A. Ginawi, J. Montstream, S. Mitra, K. Chatty, A. Joshi, K. Henderson, N. Palmer, and B. Hulse. An ESD design automation framework and tool flow for nano-scale CMOS technology. *Proceedings of the Electrical Overstress/Electrostatic Discharge (EOS/ESD) Symposium*, 2010; 91–96.

30. G.C. Tian, Y.P. Xiao, D. Connerney, T.H. Kang, A. Young, and Q. Liu. A predictive full chip dynamic ESD simulation and analysis tool for analog and mixed signal ICs. *Proceedings of the Electrical Overstress/Electrostatic Discharge (EOS/ESD) Symposium*, 2011; 285–293.

31. N. Chang, Y.L. Liao, Y.S. Li, P. Johari, and A. Sarkar. Efficient multi-domain ESD analysis and verification for large SOC designs. *Proceedings of the Electrical Overstress/Electrostatic Discharge (EOS/ESD) Symposium*, 2011; 300–306.

32. N. Trivedi, H. Gossner, H. Dhakad, B. Stein, and J. Schneider. An automated approach for verification of on-chip interconnect resistance for electrostatic discharge paths. *Proceedings of the Electrical Overstress/Electrostatic Discharge (EOS/ESD) Symposium*, 2011; 307–314.

33. H. Marquardt, H. Wagieh, E. Weidner, K. Domanski, and A. Ille. Topology-aware ESD checking: A new approach to ESD protection. *Proceedings of the Electrical Overstress/Electrostatic Discharge (EOS/ESD) Symposium*, 2012; 85–90.

34. S. Voldman. *Electrical Overstress (EOS): Devices, Circuits, and Systems*. Chichester, UK: John Wiley & Sons, Ltd, 2013.

35. S. Voldman. *ESD Basics: From Semiconductor Manufacturing to Product Use*. Chichester, UK: John Wiley & Sons, Ltd, 2012.

36. S. Voldman. *ESD: Design and Synthesis*. Chichester, UK: John Wiley & Sons, Ltd, 2011.

37. S. Voldman. Structure and method for latchup improvement using wafer via latchup guard ring. U.S. Patent No. 7,989,282, August 2, 2011.

38. S. Voldman. Structure and method for latchup improvement using wafer via latchup guard ring. U.S. Patent No. 8,390,074, March 5, 2013.

39. P. Chapman, D.S. Collins, and S. Voldman. Structure and method for latchup robustness with placement of through wafer via within CMOS circuitry. U.S. Patent No. 8,420,518, April 16, 2013.

40. S. Voldman. ESD network circuit with a through wafer via structure and a method of manufacture. U.S. Patent No. 8,232,625, July 31, 2012.

41. P. Chapman, D.S. Collins, and S. Voldman. Structure and method for latchup robustness with placement of through wafer via within CMOS circuitry. U.S. Patent No. 8,017,471, September 13, 2011.

42. P. Chapman, D.S. Collins, and S. Voldman. Structure for a latchup robust array I/O using through wafer via. U.S. Patent No. 7,855,420, December 21, 2010.

43. P. Chapman, D.S. Collins, and S. Voldman. Latchup robust array I/O using through wafer via. U.S. Patent No. 7,741,681, June 22, 2010.

44. P. Chapman, D.S. Collins, and S. Voldman. Structure for a latchup robust gate array using through wafer via. U.S. Patent No. 7,696,541, April 13, 2010.

45. P. Chapman, D.S. Collins, and S. Voldman. Latchup robust gate array using through wafer via. U.S. Patent No. 7,498,622, March 3, 2009.

7 Analog-ESD Signal Pin Co-synthesis

7.1 ANALOG SIGNAL PIN

Receiver circuits are very important in analog electrostatic discharge (ESD) design because of the ESD sensitivity of these networks [1–7]. Typically, the analog receiver circuits are the most sensitive circuits in a chip application. Receiver performance has a critical role in the semiconductor chip performance. The primary reasons for this are as follows:

- Analog receiver circuits are small in physical area.

- Analog performance requirements limit the ESD loading allowed on the receiver. MOSFET gate area, bipolar emitter area, and electrical interconnect wiring widths impact the receiver performance.

- Analog receiver inputs are electrically connected to either MOSFET gate (in a CMOS receiver) where the MOSFET gate dielectric region is the most ESD-sensitive region in RF MOSFET receiver networks. RF MOSFET gate dielectric scales with the RF performance objectives.

- Analog receiver inputs are electrically connected to the bipolar base region (in a bipolar receiver) where the bipolar transistor emitter–base junction is the most ESD-sensitive region of the bipolar transistor. The base region scales with RF performance objectives.

- Both the MOSFET gate dielectric region and the bipolar transistor base region are the more sensitive region of the structures.

- Analog receivers require low series resistance.

Analog signal pins will require ESD protection on a semiconductor chip to protect from physical damage [1–10]. Cosynthesis of the analog functional circuit and the ESD protection provides improved optimization for the analog function [10, 11]. Within a semiconductor

ESD: Analog Circuits and Design, First Edition. Steven H. Voldman.
© 2015 John Wiley & Sons, Ltd. Published 2015 by John Wiley & Sons, Ltd.

chip, the analog circuit may have multiple instances of an identical circuit. For some applications, these signal pin outputs are connected to a common circuit.

A challenge for today's analog applications is to provide differential pair receiver circuitry. The challenges today in differential pair circuitry are:

- Matching

- Low capacitance

For ESD protection, the challenges are twofold:

- Signal pin-to-rail ESD protection

- Differential pair pin-to-pin ESD protection

In some analog applications, it is important to have matching characteristics. Due to across chip linewidth variation (ACLV), spatial separation of these circuits can lead to process-induced global variations. In these implementations, global variations can also occur in the ESD networks that are electrically connected to the analog signal pins. ACLV global variations and orientation can lead to variations in both the analog circuit and ESD network. Receiver circuits require low capacitance loading for high-speed applications. Receiver circuits are typically small and hence sensitive to ESD and EOS events. With the spatial separation, signal pin-to-signal pin ESD failure is not a critical issue.

As semiconductor products are scaled, power bus and electrical connections to ESD networks are scaled in ASICs, standard cell foundry design, and memory products. In the past, bus resistance was an ESD design issue between the ESD input circuit element and the ESD power clamp element.

These issues will be discussed in the following sections, as well as solutions to improve these issues using common centroid design practices as well as in combination with parasitic elements. In this chapter, we will introduce the concept of using a common centroid for the circuit, the signal pin-to-rail ESD, and then codesign and cosynthesis of the differential pair pin-to-pin ESD protection.

7.2 ANALOG SIGNAL DIFFERENTIAL RECEIVER

In analog differential circuits, at least two signal lines are connected to a common analog circuit [3–8]. In differential circuits, the signal lines and elements within the circuit are closely spaced. Differential circuits are sensitive to local process-induced variations that can impact the matching of resistor, capacitor, and transistor elements.

7.2.1 Analog Signal CMOS Differential Receivers

Figure 7.1 shows an example of a CMOS differential pair circuit with signal pin ESD networks [3–6]. In practice, these circuits are small and hence sensitive to ESD events. ESD testing of differential circuits requires an ESD pin-to-rail test (e.g., signal pin to V_{DD}

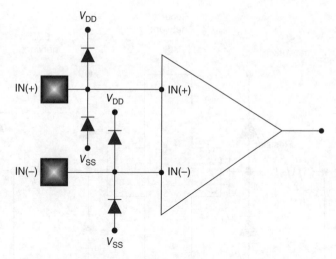

Figure 7.1   Differential input schematic with signal pin ESD.

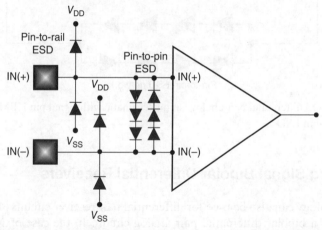

Figure 7.2   CMOS differential pair analog circuit schematic with signal pin ESD and pin-to-pin ESD.

and signal pin to V_{SS}), as well as a signal pin-to-signal pin test. In a differential pair circuit, where there is adjacency of elements connected to bond pads, ESD failure due to signal pin-to-signal pin ESD events can lead to low ESD results of the analog semiconductor chip. Figure 7.2 shows a solution for both signal pin-to-rail ESD protection and pin-to-pin ESD networks. The ESD signal pin-to-pin network must be bidirectional and symmetric.

Figure 7.3 shows a solution for both signal pin-to-rail human body model (HBM) and charged device model (CDM) ESD protection and pin-to-pin ESD networks. In analog applications with small semiconductor chips, CDM failure levels are significantly lower. In mixed-signal analog-to-digital semiconductor chips, it is a larger concern due to the charge storage of the larger semiconductor chip.

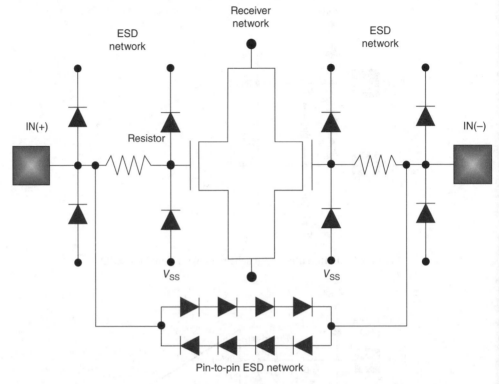

Figure 7.3 CMOS differential pair analog circuit schematic with signal pin HBM and CDM ESD and pin-to-pin ESD.

7.2.2 Analog Signal Bipolar Differential Receivers

Bipolar technology can also be used for differential pair receiver circuits [4]. Figure 7.4 is an example of a bipolar differential pair analog circuit. In the case of MOSFETs, the current does not flow through the differential pair transistors due to the MOSFET gate dielectric that is connected to the signal pins. But, in bipolar differential pair circuitry, current can flow into the metallurgical junction of the base–emitter and base–collector of the npn transistors. Receiver circuits are a common ESD-sensitive circuit in bipolar and bipolar-CMOS (BiCMOS) technology. Bipolar receiver circuits typically consist of npn bipolar transistor configured in a common emitter configuration (Figure 7.4). For bipolar receivers, the input pad is electrically connected to the base contact of the npn transistor, with the collector connected to V_{CC} either directly or through additional circuitry. The npn bipolar transistor emitter is electrically connected to V_{SS}, or through an emitter resistor element, or additional circuitry.

One of the unique problems with differential receiver networks is pin-to-pin ESD failure mechanisms. In ESD testing, we can apply an ESD pulse event to one of the two differential signal pads and use the second differential signal pad as the ground reference. In differential pair bipolar receiver networks, for positive polarity HBM ESD events,

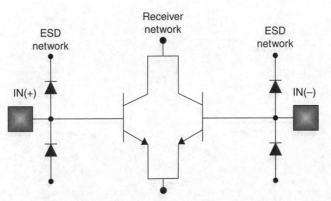

Figure 7.4 Bipolar differential pair analog circuit schematic with signal pin ESD.

as the base voltage increases, the base–emitter voltage of the first transistor increases, leading to forward biasing of the base–emitter junction. The base–emitter junction becomes forward active, leading to current flowing from the base to the emitter region. For the second npn bipolar transistor, the base–emitter region is reverse biased. As the voltage on the first signal pad increases, the base–emitter reverse-bias voltage across the second transistor base–emitter metallurgical junction increases.

In bipolar receiver networks, for positive polarity HBM ESD events, as the base voltage increases, the base–emitter voltage increases, leading to forward biasing of the base–emitter junction. The base–emitter junction becomes forward active, leading to current flowing from the base to the emitter region. Typically in bipolar receiver networks, the physical size of the emitter regions is small. When the ESD current exceeds the safe operating area (SOA), degradation effects occur in the bipolar transistor. The bipolar device degradation is observed as a change in the transconductance of the bipolar transistor. From the electrical parametrics, the unity current cutoff frequency, f_T, decreases with increased ESD current levels. From a f_T–I_C plot, the f_T magnitude decreases with ESD pulse events, leading to a decrease in the peak f_T. Avalanche breakdown occurs in the emitter–base metallurgical junction, leading to an increase in the current flowing through the emitter and base regions; this leads to thermal runaway and second bipolar breakdown in the grounded second bipolar transistor of the differential pair. Note that the degradation of the second transistor prior to the first transistor can also lead to a differential offset hampering the matching of the two sides of the differential pair. It is possible that the failure criterion is associated with an npn mismatch prior to the ESD failure of either npn device (Figure 7.5).

For a negative pulse event, the base–emitter region is reverse biased. As the voltage on the signal pad decreases, the base–emitter reverse-bias voltage across the base–emitter metallurgical junction increases. Avalanche breakdown occurs in the emitter–base metallurgical junction, leading to an increase in the current flowing through the emitter and base regions; this leads to thermal runaway and second bipolar breakdown in the bipolar transistor. The experimental results show that the negative polarity failure level has a lower magnitude compared to the positive polarity failure level.

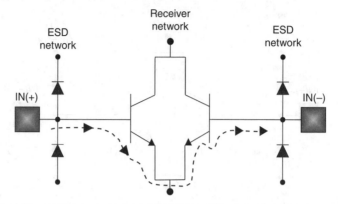

Figure 7.5   Current path for pin-to-pin test for bipolar differential pair receiver with only signal pin ESD.

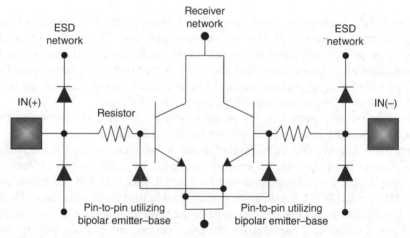

Figure 7.6   Bipolar differential pair analog circuit schematic with signal pin ESD and emitter–base ESD device.

An ESD design solution used to provide improved ESD results in a differential pair bipolar receiver network is to place a p–n diode element in parallel with the npn bipolar transistor emitter–base junction (Figure 7.6). Using a parallel element, the p–n junction is placed such that the anode is electrically connected to the npn emitter and the cathode to the base region. In this fashion, an alternate forward-bias current path is established between both sides of the differential pair. Figure 7.7 shows the current path through the differential pair emitter–base and the additional diode element.

An alternative method is to introduce multi-emitters within the base region of the npn transistors. Multi-emitter bipolar transistors were utilized in bipolar technology in static RAM (SRAM) memory cells. Figure 7.8 provides an example of a multi-emitter bipolar transistor, where a larger emitter is utilized for the ESD protection scheme [10].

Another method is to introduce a back-to-back diode string between both sides of the differential pair. Figure 7.9 shows a bipolar differential pair with an ESD diode string

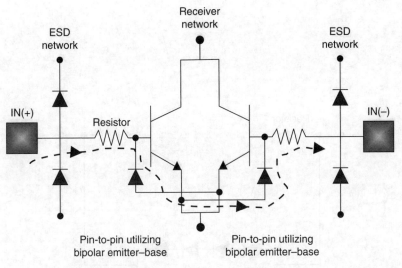

Figure 7.7 Current path in bipolar differential pair analog circuit schematic with signal pin ESD and emitter–base ESD device.

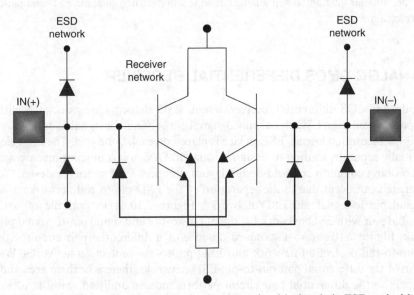

Figure 7.8 Bipolar differential pair analog circuit schematic with signal pin ESD and with multi-emitter emitter–base ESD device.

between the two sides of the differential pair. This has the advantage of allowing a higher current between both sides of the differential pair and avoiding emitter–base degradation but has the disadvantage of asymmetry matching and capacitance loading performance degradation.

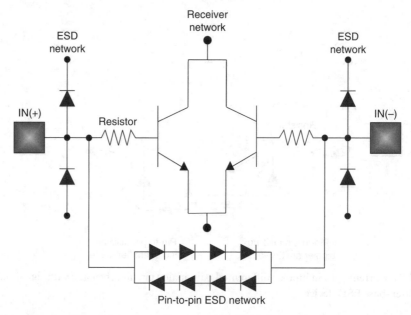

Figure 7.9   Bipolar differential pair analog circuit schematic with signal pin ESD and pin-to-pin ESD protection.

7.3 ANALOG CMOS DIFFERENTIAL RECEIVER

In standard CMOS differential receiver design, ESD elements are associated with each signal pad, as well as an ESD signal pin-to-signal pin between the signal pads (Figure 7.10). In the implementation layout, ESD device is placed on each bond pad. The ESD elements are spatially separate, leading to design mismatch. Design layout solutions are achieved by establishing common centroid design layout practices used in analog design. Sources of mismatch can occur due to the separation of the ESD pin-to-rail networks as well as the signal pin-to-signal pin ESD networks. Figure 7.10 is an example of a CMOS differential pair with ESD network for signal pin-to-rail and signal pin-to-signal pin ESD network. Figure 7.10 shows a standard practice of a differential pair circuit with both ESD pin-to-rail protection network and ESD pin-to-pin protection networks. With the addition of the differential pair pin-to-pin ESD networks, there is both an area and loading impact to the differential pair circuit performance. In addition, without a common centroid implementation, it will introduce a mismatch [4–7].

7.3.1 Analog Differential Receiver Capacitance Loading

A second issue, given that either diode, MOSFET, or SCR ESD networks are used, the loading capacitance of the ESD network impacts differential pair receiver performance. In the case of a dual-diode ESD network, two additional diode capacitances are added to

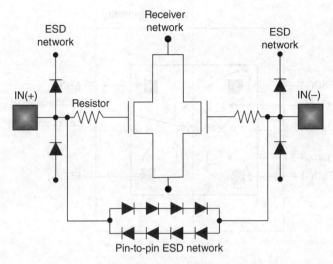

Figure 7.10 Standard practice for ESD pin-to-pin network for differential pair circuits.

the differential pair network for both IN(+) and IN(−). This common centroid concept is typically not extended to ESD networks.

7.3.2 Analog Differential Receiver ESD Mismatch

ESD signal pin-to-rail protection networks impact the mismatch and loading capacitance of the differential pair circuitry. The first critical issue is the mismatch introduced between the two sides of the differential pair from the ESD structure itself. With the addition of the ESD network on the IN(+) and second ESD network on IN(−), a mismatch occurs since both ESD networks are spatially separated. Hence, the spatial separation of the two separate ESD networks can lead to functional implications in itself (Figure 7.11).

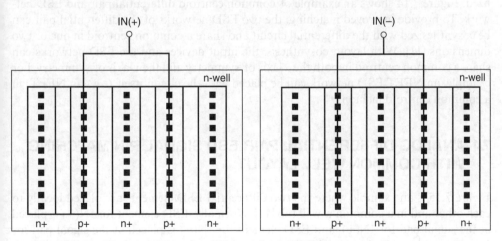

Figure 7.11 Placement of analog differential pair ESD circuits without common well.

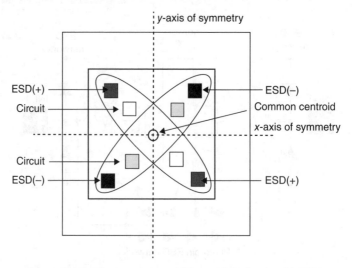

Figure 7.12   Common centroid for analog circuit with ESD.

Symmetry is important for minimizing design variation for both circuits and ESD networks [9]. A metric to define symmetry is by establishing an axis of symmetry. Figure 7.12 demonstrates a common centroid in analog differential pair networks [9].

Common centroid design introduces four rules: (1) coincidence, (2) symmetry, (3) dispersion, and (4) compactness [9]. To minimize variations, symmetry can be evaluated in one dimension or both dimensions. For example, an axis of symmetry can be defined in the x-axis, and a second axis of symmetry can be defined in the y-axis. From these axes of symmetry, a common centroid can be established. Common centroid design is a process used in analog design practices. Figure 7.13 is an example of a common centroid differential pair with x- and y-symmetry [11].

For the case of grounded-gate MOSFETs, common centroid design can also be utilized. Figure 7.14 shows an example of common centroid differential pair and ESD network. To provide improved matching, the two ESD networks of the differential pair can be cosynthesized with the differential circuit and share a common centroid in one or two dimensions [11]. With layout cosynthesis, the input devices and the ESD networks can share a common centroid in both the x-axis of symmetry and the y-axis of symmetry. For example, an NFET ESD network can be placed with the circuit itself (e.g., an NFET) in a common centroid fashion.

7.4 ANALOG DIFFERENTIAL PAIR ESD SIGNAL PIN MATCHING WITH COMMON WELL LAYOUT

Figure 7.15 is an example where the two differential ESD elements are placed locally to each other and in a common n-well tub [11]. Normally, ESD element for IN(+) is placed at the signal pad for IN(+), and ESD element for IN(−) is placed at the signal pad for IN(−). In this example, the ESD elements are placed into a common shared region.

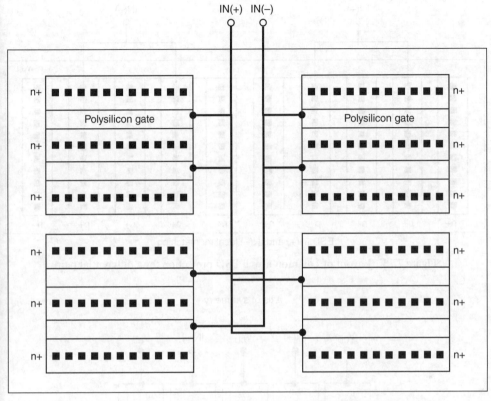

Figure 7.13   Common centroid for analog circuit with *x*- and *y*-symmetry.

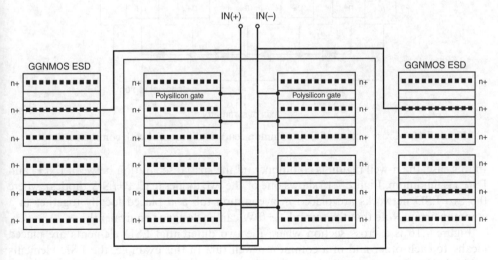

Figure 7.14   Common centroid differential pair with *x*- and *y*-symmetry for analog circuit with GGNMOS.

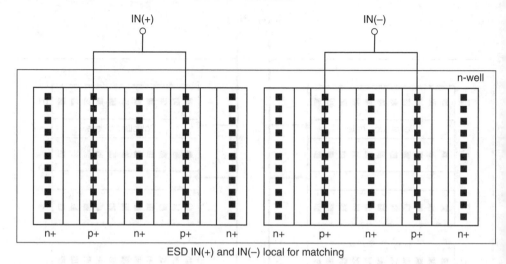

Figure 7.15   Layout of common n-well ESD protection for a differential pair.

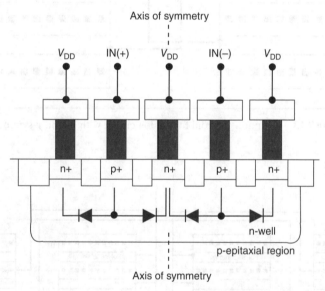

Figure 7.16   Cross section of ESD common centroid differential with common well tub.

In this case, the n-well region is common, and the spatial separation between IN(+) and IN(−) is minimum. Figure 7.15 shows one of the two diodes (e.g., P+/NW diodes), where the two ESD networks are placed in a common tub and placed locally together in a common array. Note that the second N+/PW can be designed equivalently.

Figure 7.16 is a cross section where the two differential ESD elements are placed locally to each other and in a common n-well tub. In this example, the ESD elements are placed into a common shared region. In this case, the n-well region is common, and the spatial separation between IN(+) and IN(−) is minimum. Figure 7.16 shows one of the two diodes (e.g., P+/NW diodes), where the two ESD networks are placed in a common tub and placed locally together in a common array. Note that the second N+/PW can be designed equivalently [11].

7.5 ANALOG DIFFERENTIAL PAIR COMMON CENTROID DESIGN LAYOUT: SIGNAL PIN-TO-SIGNAL PIN AND PARASITIC ESD ELEMENTS

ESD signal pin-to-signal pin protection networks are required to provide ESD protection from the two pins within a differential pair circuit. Typically in analog applications (e.g., CMOS and bipolar), the differential pair pins are the most sensitive pins in a given semiconductor chip. It has been also shown that in "signal pin to all other signal pins (reference ground)," the failure mechanism occurs between the two pins of the differential pair [11].

For CMOS differential pair, there are two solutions for establishing the signal pin-to-signal pin ESD networks. Presently, the established conventional method is to utilize a bidirectional ESD network between the IN(+) and IN(−). Typically, it is not a common centroid implementation.

A novel method is to introduce a common centroid implementation, where one ESD array serves both sides of the differential pair network. In this fashion, a common centroid design can be achieved. By taking the next step of alternating the fingers of the one ESD array, the parasitic elements between the two sides can be used.

For example, a lateral parasitic pnp can be formed between adjacent fingers of the differential pair ESD network p+/n-well diode. Additionally, a lateral parasitic npn can be used between adjacent fingers of the differential pair ESD network n+/p-well diode (Figure 7.17).

This achieves multiple objectives:

- Common centroid design with improved matching

- No additional capacitance load on the differential pair receiver network

- No additional area for an additional ESD network

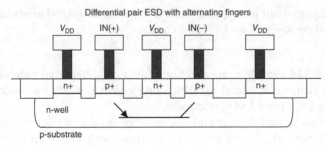

Figure 7.17 Differential pair interdigitated common centroid design and utilization of parasitic elements for signal pin-to-signal pin ESD protection.

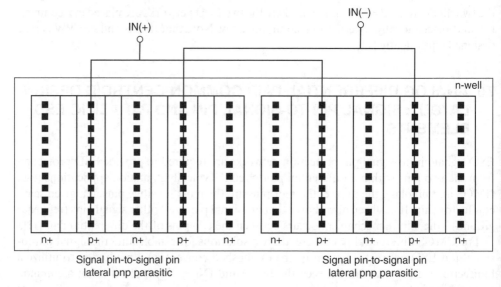

Figure 7.18   Differential pair interdigitated common centroid design and utilization of parasitic elements for signal pin-to-signal pin ESD protection.

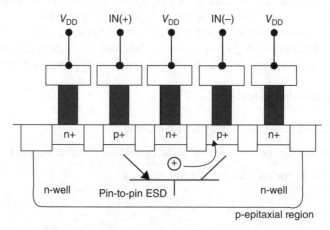

Figure 7.19   Differential pair interdigitated common centroid design and utilization of parasitic elements for signal pin-to-signal pin ESD protection.

Figures 7.18 and 7.19 provide a cross section utilizing a differential pair with a parasitic pin-to-pin ESD element. A lateral parasitic pnp element is used to provide ESD protection between the IN(+) and IN(−) electrodes.

Figure 7.20 shows a high-level circuit schematic of the differential pair with a parasitic pin-to-pin ESD element. It also shows an example with the parasitic pnp. A lateral parasitic pnp element is used to provide ESD protection between the IN(+) and IN(−) electrodes [11].

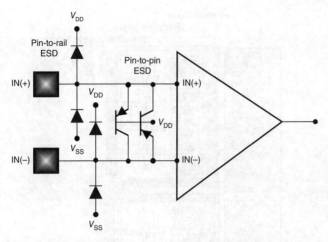

Figure 7.20   Circuit schematic of differential pair with interdigitated pin-to-pin ESD devices.

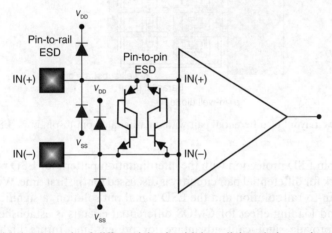

Figure 7.21   Circuit schematic of differential pair with interdigitated pin-to-pin pnpn ESD devices.

Figure 7.21 shows an example with the parasitic pnpn [11]. A lateral parasitic pnpn element is used to provide ESD protection between the IN(+) and IN(−) electrodes. Figure 7.22 shows an example layout using the parasitic pnpn.

7.6 CLOSING COMMENTS AND SUMMARY

This chapter addressed analog and ESD signal pin cosynthesis that introduces usage of interdigitated layout and common centroid concepts to provide ideal matching, low capacitance, and small area in differential signal pins. Using interdigitated designs, the parasitic elements are utilized for signal pin-to-signal pin ESD protection. In conclusion, the issue of common centroid design of ESD protection networks which integrates signal

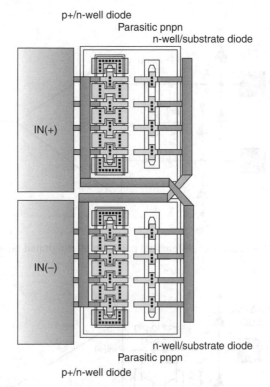

Figure 7.22   Layout of differential pair with interdigitated pin-to-pin pnpn ESD devices.

pin-to-signal pin ESD protection with the interdigitation pattern for ESD pin-to-rail protection network for differential pair circuitry is discussed for the first time. With integration of the ESD pin-to-rail solution and the ESD signal pin solution, a significant reduction of the area and loading effect for CMOS differential circuits is established. This novel concept will provide significant advantage for present and future high-performance analog and RF design for matching, area reduction, and performance advantages.

Chapter 8 focuses on floor planning in analog and mixed-signal chip design. Core and peripheral I/O design will be addressed associated with analog applications. Integration of analog and digital ESD networks within the semiconductor chip floor plan is quantified. ESD power clamp placement and integration for independent and master/slave implementation systems were shown. Examples of rectangular and octagonal bond and integration of circular and octagonal ESD elements are shown. The chapter closes with the current and the future issues for 2.5-D and 3-D systems.

REFERENCES

1. V. Vashchenko and A. Shibkov. *ESD Design for Analog Circuits.* New York: Springer, 2010.
2. S. Voldman. *ESD: Physics and Devices.* Chichester, UK: John Wiley & Sons, Ltd, 2004.
3. S. Voldman. *ESD: Circuits and Devices.* Chichester, UK: John Wiley & Sons, Ltd, 2005.

4. S. Voldman. *ESD: RF Circuits and Technology*. Chichester, UK: John Wiley & Sons, Ltd, 2006.

5. S. Voldman. *ESD: Failure Mechanisms and Models*. Chichester, UK: John Wiley & Sons, Ltd, 2009.

6. S. Voldman. *ESD: Design and Synthesis*. Chichester, UK: John Wiley & Sons, Ltd, 2011.

7. S. Voldman. *ESD Basics: From Semiconductor Manufacturing to Product Use*. Chichester, UK: John Wiley & Sons, Ltd, 2012.

8. S. Voldman. *Electrical Overstress (EOS): Devices, Circuits, and Systems*. Chichester, UK: John Wiley & Sons, Ltd, 2013.

9. A. Hastings. *The Art of Analog Layout*. Upper Saddle River, NJ: Prentice Hall, 2006.

10. S. Voldman. Dual emitter transistor with ESD protection. U.S. Patent No. 6,731,488, May 4, 2004.

11. S. Voldman. Common centroid differential pair signal pin-to-signal pin analog ESD design. *Proceedings of the IEEE 11th International Conference on Solid-State and Integrated Circuit Technology (ICSICT)*, 2012.

8 Analog and ESD Circuit Integration

8.1 ANALOG AND POWER TECHNOLOGY AND ESD CIRCUIT INTEGRATION

In analog circuitry and power applications, a large number of electrostatic discharge (ESD) networks are needed due to the breadth of applications, application voltages, and technology requirements [1–11]. ESD networks must be suitable for the application voltages ranging from 1.5 V to ultrahigh-voltage (UHV) conditions of 600–800 V. In this chapter, the discussion will go into greater depth on the needs of an analog and power technology, with some focus on the issues for analog power ESD networks.

8.1.1 Analog ESD: Isolated and Nonisolated Designs

In analog circuitry, the ability to isolate the network from the substrate is necessary for voltage isolation, noise, or latchup [11]. In LDMOS and bipolar-CMOS-DMOS (BCD) technology, transistor elements can be placed in the substrate or in an isolation well structure. ESD protection networks can be constructed with both isolated and nonisolated designs. For ESD designs that are isolated, the isolation voltage is the maximum voltage that the element can be used for a given power supply application voltage.

8.1.2 Integrated Body Ties

In high-voltage (HV) applications from 5 to 120 V, an important issue is whether a device has a "body tie" electrically connecting a source to the LDMOS body [11]. In some ESD networks, it may be required to not connect the transistor body to the source but

ESD: Analog Circuits and Design, First Edition. Steven H. Voldman.
© 2015 John Wiley & Sons, Ltd. Published 2015 by John Wiley & Sons, Ltd.

have the ability to isolate them. In networks that require "stacking" of elements, it may require the electrical separation of the source and the "body" of an LDMOS transistor.

8.1.3 Self-Protecting versus Non-Self-Protecting Designs

In HV applications from 5 to 120 V, analog circuitry can be designed to be "self-protecting" without additional ESD networks [11]. For power applications, the physical width of "pull-up" and "pull-down" transistors can be very large. As a result, an ESD network may not be necessary. Self-protecting analog circuits may require specific layout practices that allow for more ESD or electrical overstress (EOS) robustness than standard LDMOS or DeMOS devices.

8.2 ESD INPUT CIRCUITS

ESD protection of input circuit is critical due to the ESD sensitivity of input circuitry. Input circuits are typically small circuits. In the following sections, analog input circuits for both low voltage (LV) and HV are discussed.

8.2.1 Analog Input Circuit Protection

Analog input protection is critical due to the sensitivity of input circuitry. In addition, for power electronics, the issues of application voltages, isolation voltage, ESD "turn-on voltages," and ESD device breakdown voltages are important to have proper cosynthesis of analog circuits and ESD devices. This section will introduce additional analog ESD networks for power devices.

8.2.2 High-Voltage Analog Input Circuit Protection

In HV applications from 5 to 120 V, HV analog inputs will require usage of ESD elements whose breakdown voltage to the isolation or substrate is above the application voltage. Whereas LV CMOS can utilize CMOS diode-based elements, this is not possible when the n-well-to-substrate breakdown voltage is below the application voltage. As a result, LDMOS transistors can be used in LDMOS and BCD technology for analog input ESD networks. Figure 8.1 shows an example of a single-stage LDMOS-based input network. The analog ESD network uses a p-channel LDMOS device between analog input and the high-voltage power supply voltage V_{CCHV} and an n-channel LDMOS device between analog input and ground (V_{ss}). In these circuits, the LDMOS gate breakdown voltage must be suitable for the application voltage, as well as the isolation breakdown voltage through the signal swing of the application, and relative to the power supply voltage V_{CCHV}.

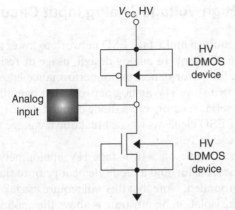

Figure 8.1   Analog input protection with LDMOS transistors.

8.2.3 Analog Input High-Voltage Grounded-Gate NMOS (GGNMOS)

In analog design, HV inputs utilize a commonly used voltage triggered ESD network, which is, which is the HV grounded-gate NMOS (GGNMOS) network. Figure 8.2 is an example of an HV GGNMOS network used for analog and power applications. For analog and power applications, the HV GGNMOS transistor must have a high dielectric breakdown voltage to avoid gate-to-drain breakdown. LDMOS transistors that have breakdown voltage tolerance are commonly used. For power applications, the GGNMOS device utilized must also have an isolation voltage above the application voltage. In many power applications, the "body" can float to avoid electrical breakdown in ESD protection networks.

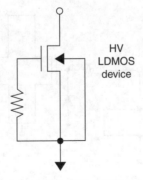

Figure 8.2   Analog input protection with LDMOS transistors.

8.2.4 Two-Stage High-Voltage Analog Input Circuit Protection

HV analog inputs will require a multistage ESD network to lower the voltage at the input pin during an ESD or EOS event. In analog design, usage of resistor elements in series with inputs is acceptable for performance- or nonperformance-based networks. Figure 8.3 shows an example of a two-stage HV analog network that contains a PDMOS element, an NDMOS element, a series resistor, and a diode element prior to the circuit. Note that this will require usage of ESD elements whose breakdown voltage to the isolation or substrate is above the application voltage.

Figure 8.4 shows an example of a two-stage HV analog network that contains an NDMOS element, a series resistor, and a diode element prior to the circuit. In this implementation, the body is grounded. Note that this will require usage of ESD elements whose breakdown voltage to the isolation or substrate is above the application voltage.

Figure 8.5 shows an example of a two-stage HV analog network that contains an NDMOS element, a series resistor, and a diode element with the body floating. This will allow for usage of the LDMOS transistor without isolation breakdown.

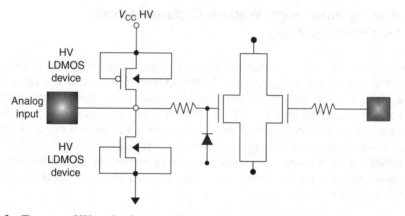

Figure 8.3 Two-stage HV analog input protection using LDMOS transistors, resistor, and diode.

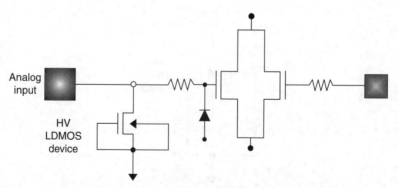

Figure 8.4 Two-stage HV analog input protection using GGNMOS LDMOS transistor, resistor, and diode.

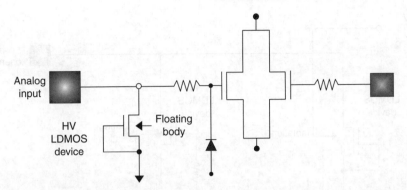

Figure 8.5   Two-stage HV analog input protection using floating-body LDMOS transistor.

8.3 ANALOG ESD OUTPUT CIRCUITS

In HV applications, analog ESD solutions for output networks used can be utilized for output drivers.

8.3.1 Analog ESD Output Networks and Distinctions

In HV applications, analog ESD solutions for output networks used for receiver networks can be utilized for output drivers. Some distinctions for output driver networks are as follows:

- Self-protection is an option for large output driver elements.
- Some topologies require ESD networks.
- Isolation voltage conditions must avoid breakdown or limiting the signal swing of the output network.

In HV applications from 5 to 120 V, analog circuitry can be designed to be "self-protecting" without additional ESD networks. For power applications, the physical width of "pull-up" and "pull-down" transistors can be very large. As a result, an ESD network may not be necessary. Self-protecting analog circuits may require specific layout practices that allow for more ESD or EOS robustness than standard LDMOS or DeMOS devices.

In some HV applications, some topologies have only a "pull-down" network and do not allow for a "pull-up" element. An example of this is an "open-drain" topology.

8.3.2 Analog Open-Drain ESD Output Networks

An analog open-drain output network only has a "pull-down" network and does not allow for a "pull-up" element. Figure 8.6 is an example of an HV open-drain network schematic with ESD protection. An HV LDMOS device is electrically connected to the output voltage. The isolation of the LDMOS device is electrically connected to the HV

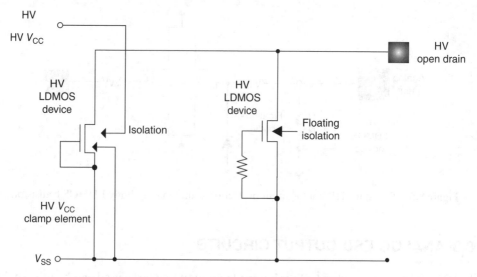

Figure 8.6 HV V_{CC} open-drain output network.

V_{CC} power supply, whereas the body region is electrically connected to ground V_{SS}. An HV LDMOS network is also placed between the open-drain output and the ground. In this element, the isolation is left floating.

8.4 ANALOG ESD GROUND-TO-GROUND NETWORKS

In analog and digital-to-analog mixed-signal technology, electrical connectivity and isolation can be established through the ground rails with ESD ground-to-ground networks [15, 16]. These networks can have the following features:

- Unidirectional

- Bidirectional

- Symmetric

- Asymmetric

- Single elements or series (e.g., stack) of elements

In product implementations, the choice of the design is a function of the trade-off of ESD connectivity versus noise isolation. Unidirectionality can prevent current flow or current injection in one direction. Bidirectionality allows flow in both directions between the power rails. Using symmetric versus asymmetric designs allows preferential isolation and capacitive coupling.

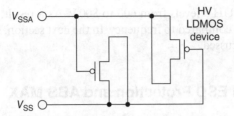

Figure 8.7 HV ground-to-ground diode-configured LDMOS.

8.4.1 Back-to-Back CMOS Diode String

In CMOS technology, LDMOS technology, and BCD technology, ground-to-ground ESD networks can consist of CMOS-based element. CMOS elements can be p+/n-well diodes, gated diodes, or diode-configured MOSFETs. A disadvantage of using an LV CMOS MOSFET is electrical failure of the MOSFET gate structure. Additionally, the devices may need to be electrically isolated for noise and latchup. CMOS diode structures can contain n-wells which collect minority carrier from the substrate.

Using triple well technology, isolation can be achieved to minimize the problem of noise injection by separating from the substrate region. The triple well diode structures will need a power supply voltage electrically connecting to the isolating structures; the issue with this is that the minority carrier will be collected by the analog V_{DD} (AV_{DD}).

8.4.2 HV GGNMOS Diode-Configured Ground-to-Ground Network

In LDMOS technology and BCD technology, ground-to-ground ESD networks can consist of LDMOS-based elements instead of CMOS-based diode elements. Figure 8.7 shows an example of a bidirectional symmetric back-to-back network. One of the advantages of using an LDMOS device is that LDMOS technology is designed for power, whereas CMOS technology is designed for performance.

8.5 ESD POWER CLAMPS

ESD power clamps between power supply and ground connections are commonly used in all domains of a semiconductor chip. In this section, HV ESD power clamps will be first discussed, followed by discussion of LV analog and digital domain ESD power clamps [13–18].

8.5.1 ESD Power Clamp Issues for the High-Voltage Domain

In analog design, voltage-triggered networks are commonly used for power applications. In an analog design, there are many power supply requirements. Additionally, the range of power supply voltages is significant. The power supply voltage range can extend from

LV CMOS at 1.5 V to UHV levels from 600 to 800 V. As a result, voltage conditions are a more significant issue compared to frequency. In the next section, some HV ESD network and integration are discussed.

8.5.2 HV Domain ESD Protection and ABS MAX

In HV applications, the input power to HV analog sections may be as high as the LDMOS absolute maximum drain voltage rating. As a result, a concern exists of the choice of ESD network to be compliant with the isolation voltages.

8.5.3 HV Domain V_{IN} or V_{CC} Input

Figure 8.8 shows an example where the HV domain has a V_{IN} or V_{CC} input that requires ESD protection. It also shows an example embodiment using a GGNMOS stack network, with a floating body, as well as a GGNMOS return diode element across the power supplies of an amplifier network. The choice of the ESD network will be required to be compliant with the isolation voltages.

8.5.4 HV Grounded-Gate NMOS (GGNMOS)

In analog design, a voltage-triggered ESD network commonly used is the HV GGNMOS network [11]. Figure 8.9 is an example of an HV GGNMOS network used for analog and power applications. For analog and power applications, the HV GGNMOS transistor must have a high dielectric breakdown voltage to avoid gate-to-drain breakdown. LDMOS transistors that have breakdown voltage tolerance are commonly used. For power applications,

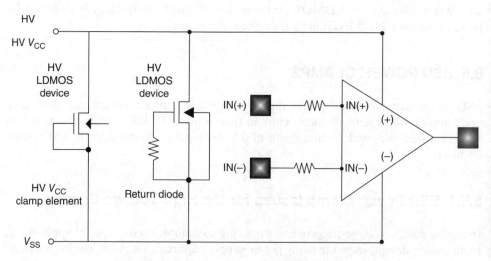

Figure 8.8   HV V_{CC} power input protection scheme.

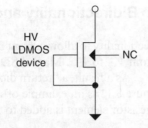

Figure 8.9   HV ESD GGNMOS.

the GGNMOS device utilized must also have an isolation voltage above the application voltage. In many power applications, the "body" can float to avoid electrical breakdown in ESD protection networks.

8.5.5 HV Series Cascode ESD Network

In analog design, a voltage-triggered ESD network commonly used is the HV GGNMOS network. For analog and power applications, the HV GGNMOS transistor must have a high dielectric breakdown voltage to avoid gate-to-drain breakdown. LDMOS transistors that have breakdown voltage tolerance are commonly used. For power applications, the GGNMOS device utilized must also have an isolation voltage above the application voltage. In many power applications, the "body" can float to avoid electrical breakdown in ESD protection networks. Figure 8.10 is an example of an HV GGNMOS network where elements are in a series cascode configuration for HV analog and power applications [11].

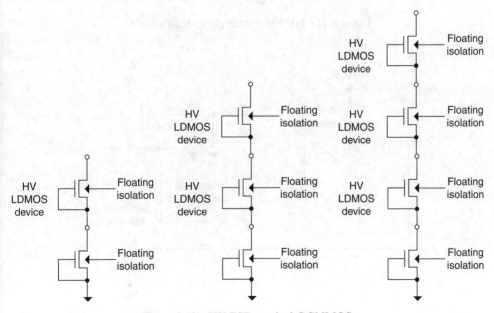

Figure 8.10   HV ESD stacked GGNMOS.

8.5.6 ESD Power Clamp Bidirectionality and Return Diodes

ESD power clamps require bidirectionality to allow current flow in a first direction from V_{CC} to V_{SS} and a second direction from V_{SS} to V_{CC}. Many ESD power clamps are designed to allow current flow from V_{CC} to V_{SS}. As a result, a "return diode" is added to provide current flow in the reverse direction. Figure 8.11 is an example of a "return diode" ESD network that uses an HV GGNMOS. A resistor element is added to the LDMOS gate electrode.

8.5.7 Alternative Solutions: LDO Current Limits

Alternative solutions exist with the use of voltage and current limiting function internal or external to the circuit function. Current limiting function can be added to the circuit network to control the electrical overcurrent (EOC) issues (Figure 8.12).

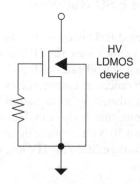

Figure 8.11   HV ESD power clamp return diode.

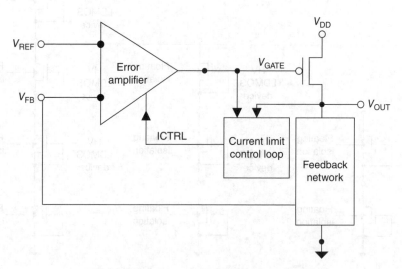

Figure 8.12   Alternative LDO current limit.

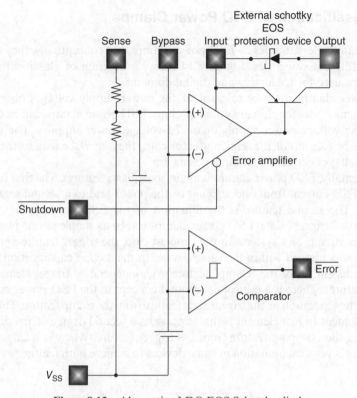

Figure 8.13   Alternative LDO EOS Schottky diode.

8.5.8 Alternative Solutions: External EOS Diode

Alternative solutions exist with the use of voltage and current limit external to the circuit function. Some ESD and EOS solutions exist on-chip and some off-chip [12]. Figure 8.13 shows an example of an EOS Schottky diode external to the semiconductor chip.

8.6 ESD POWER CLAMPS FOR LOW-VOLTAGE DIGITAL AND ANALOG DOMAIN

For digital applications, ESD power clamp usage began in the mid-1990s and today is a common practice of semiconductor chip design and ESD design synthesis [15–18]. In the later 1990s to early 2000 period, these began to be implemented into analog applications. Development of ESD power clamps and the synthesis into the semiconductor chip architecture are part of the ESD design discipline and an essential component of the art of ESD design. In this section of the chapter, the focus on the classification of the ESD power clamps, key design parameters, the ESD power clamp design window, trigger elements, clamp devices, and issues and problems with ESD power clamp will be discussed.

8.6.1 Classification of ESD Power Clamps

There are many different types of ESD power clamps, but conceptually, they can be classified into different categories. Figure 8.14 shows a diagram of classification of ESD power clamps used for LV analog and digital domains.

ESD power clamps must be tolerant of the power supply voltages observed in the functional semiconductor chip or system of chips. ESD power clamps can be constructed for the native voltage power supply or mixed-voltage power supplies. The ESD power clamps must be tolerant of the semiconductor chips they interface with or the number of power rail voltages contained within a given chip.

Fundamentally, ESD power clamps contain some basic features. The first feature is the transfer of ESD current from one segment of the power grid to a second segment of the power grid. The second feature is the initiation of the ESD power clamp, commonly referred to as a "trigger" state. ESD power clamps can be as simple as one physical device or a complex circuit, or a system. In the simplest case, the trigger feature and the clamp feature can be contained within the same device. In the second classification, the trigger element is independent of the "clamp" feature (e.g., independent trigger element from the clamping feature). There is a critical conceptual concept in the ESD power clamp design synthesis in the separation of the trigger state feature from the clamp feature. The advantage of an independent trigger element is that it provides a second degree of freedom with the separation of the clamping feature from the trigger feature. Whereas a single integrated fashion, there is physical limitation in some devices to achieve both features as desired.

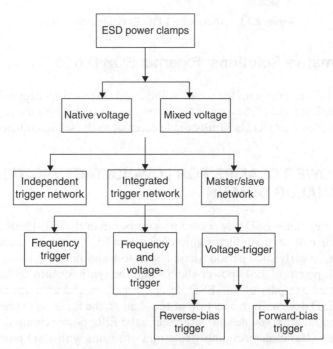

Figure 8.14 Classes of ESD power clamps.

In the third classification, the ESD power clamp is a system of ESD power clamps, with one trigger element for a system of clamp elements, which will be referred to as a "master/slave" architecture. A master/slave system allows integration of a single trigger element but allows distribution of the elements in the chip system.

In the ESD power clamp "trigger feature," there are many different solutions used for ESD power clamps, but again, they can be simply stated as classifications of trigger elements.

ESD power clamps can have trigger feature that responds to the ESD pulse. The "trigger network" is responsive to a given frequency or transient phenomena. This class of trigger networks will be referred to as "frequency triggering." Frequency trigger can contain elements that are frequency dependent, such as resistors, capacitors, and inductors [3, 4, 6]. Frequency-triggered networks respond in the frequency domain. ESD trigger elements can also be networks that do not respond in the frequency domain. These ESD trigger networks can also be initiated by overvoltage or overcurrent condition. A class of ESD trigger networks are voltage-triggered elements. Voltage-triggered elements can be initiated in a forward-bias or reverse-bias state of operation.

In this classification, there are additional features that have been added to address other characteristics. Some of these features are as follows:

- Ramping of the power supplies (e.g., power-up or power-down)

- Sequencing of power supplies

- False triggering from system events

- ESD testing precharging phenomena

- ESD testing "trailing pulse" phenomena

8.6.2 ESD Power Clamp: Key Design Parameters

In ESD design synthesis of ESD power clamps, there are key design parameters in the decision of what type of circuit to utilize. The following is a list of key parameters in the ESD design process of ESD power clamps [18]:

- ESD power clamp physical area

- ESD power clamp width

- ESD power clamp current per unit of width metric (A/um)

- ESD power clamp "on-resistance"

- ESD power clamp voltage tolerance

- ESD power clamp latchup robustness

- ESD power clamp false triggering immunity

- ESD power clamp IEC 61000-4-2 responsiveness

- ESD power clamp leakage current

- ESD power clamp capacitance loading
- ESD power clamp frequency response window
- ESD power clamp trigger voltage or current
- ESD return diode

These features and aspects of ESD power clamps will be discussed. These ESD power clamps can be made of diodes, bipolar transistors, MOSFETs, silicon-controlled rectifiers, and LDMOS transistors.

8.6.3 Design Synthesis of ESD Power Clamps

The ESD power clamp "trigger feature" is critical to initiate the ESD power clamp. ESD power clamps can have trigger feature that responds to the ESD pulse through either transient response or voltage levels. The following sections will focus on two major classes of "trigger networks."

8.6.4 Transient Response Frequency Trigger Element and the ESD Frequency Window

In ESD power clamps, the ESD power clamp trigger element can be a frequency-triggered network or transient response trigger element [18]. Transient response trigger elements are designed to respond to the ESD events. This class of trigger networks will be referred to as "frequency triggering." Frequency-triggered networks respond in the frequency domain. The frequency trigger can contain elements that are frequency dependent, such as resistors, capacitors, and inductors, in resistor–capacitor (RC), LC, or RLC configurations. In ESD power clamps, the most widely used and most popular is the RC network. The RC-triggered network is also known as "RC discriminator" network, due to it providing frequency selection in the ESD power clamp frequency domain [14, 15]. By providing a separate RC filter network, the frequency response of the trigger network will not be dependent on the inherent native frequency response of a semiconductor device and can be "tuned" to the desired frequency. In the majority of applications, the RC discriminator network is tuned to be responsive to the human body model (HBM) and machine model (MM) pulse events. One of the key advantages of frequency-triggered ESD clamps is that it is a function of the transient or rising edge, not the voltage level of the power grid.

In the frequency-triggered network, the resistor and capacitor elements can be passive or active semiconductor elements. The choice of what element to use is a function of the technology, area utilization, voltage tolerance, and device responsiveness. The resistor used for the RC network can be the following:

- Polysilicon resistor element
- Diffused resistor element

- "On" n-channel transistor element
- "On" p-channel transistor element

The capacitor element typically used for the RC network is as follows:

- MOS capacitor
- MIM capacitor

8.6.5 ESD Power Clamp Frequency Design Window

Figure 8.15 shows the ESD power clamp frequency window [18]. It provides a frequency plot highlighting the typical frequency of ESD events, overlaying the typical design point for ESD power clamps. Typically, ESD power clamps are designed to respond to the HBM and MM events (Figure 8.16). ESD power clamps are not designed to respond to charged device model (CDM) events. In addition, the ESD power clamps are not to be initiated by the power-up and power-down of the semiconductor chip or system. The ESD power clamps are not to be initiated by system events, leading to "false triggering."

As a result, there is a defined frequency window that is acceptable for ESD power clamps and the frequency range of these networks.

8.6.6 Design Synthesis of ESD Power Clamp: Voltage-Triggered ESD Trigger Elements

In ESD power clamps, the ESD power clamp trigger element can be a current- or voltage-triggered ESD network [13–18]. Voltage-triggered elements are designed to respond to the ESD events when the voltage exceeds the trigger condition. These ESD power clamps

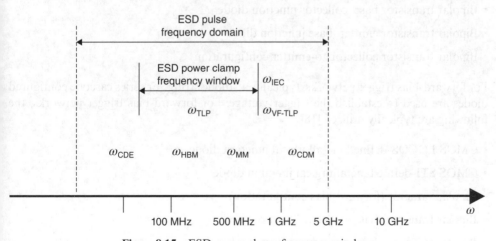

Figure 8.15   ESD power clamp frequency window.

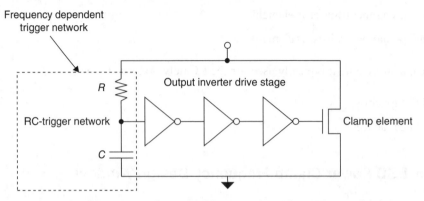

Figure 8.16 Example of frequency-triggered ESD power clamp for analog applications highlighting the trigger network.

will turn on when the voltage exceeds the trigger state. As a result, it is not dependent on the frequency of the transient event. As a result, this turns "on" the circuitry independent if it is an ESD event, EOS event, or any overvoltage or overcurrent state. These ESD power clamps are not to be initiated by the power-up and power-down of the semiconductor chip or system except when they are in an overvoltage state.

ESD voltage-triggered elements can be either forward-bias or reverse-bias elements or circuits [18]. For reverse-bias trigger networks, the following are typically utilized:

- Zener breakdown diode

- Polysilicon diode

- CMOS LOCOS-defined metallurgical junction diode

- CMOS shallow trench isolation (STI)-defined metallurgical junction diode

- Bipolar transistor collector-to-substrate junction diode

- Bipolar transistor base–collector junction diode

- Bipolar transistor emitter–base junction diode

- Bipolar transistor collector-to-emitter configuration

For forward-bias trigger networks, typically, a "diode string" or series cascode configured diodes are used to establish the trigger voltage. For forward-bias trigger networks, the following are typically utilized [18]:

- CMOS LOCOS-defined metallurgical junction diode

- CMOS STI-defined metallurgical junction diode

- Bipolar varactor (forward-bias configuration)

- Bipolar transistor base–collector junction diode

- Bipolar transistor base–emitter junction diode

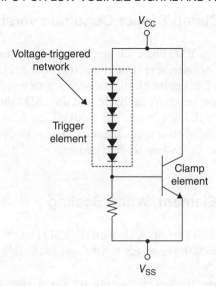

Figure 8.17 Example of voltage-triggered ESD power clamp highlighting the trigger network.

In some applications, to achieve the desired trigger voltage, the forward-bias elements can be combined with reverse-bias elements. By using the forward-bias trigger elements in series with the reverse-bias trigger element, higher trigger voltage states are achieved (Figure 8.17).

8.6.7 Design Synthesis of ESD Power Clamp: The ESD Power Clamp Shunting Element

For ESD protection power clamps, the two basic functions are the trigger network and the "shunt" network (e.g., also referred to as "clamp element") [18]. The role of the shunt element is to provide a current path in the alternative current loop to discharge the ESD current. For the effectiveness of the ESD power clamp, there are a few desired features of the ESD clamp element:

- **Low impedance:** Provide a low impedance path (e.g., a low "on-resistance")
- **ESD robustness:** Provide an ESD robust solution (e.g., discharge the ESD current without failure below the desired ESD specification)
- **Scalable:** Scalable element with physical size (e.g., width, length, perimeter, or area)

There are some additional desired characteristics of the ESD power clamp "shunt element." These consist of the following:

- ESD power clamp trigger condition versus ESD power clamp shunt failure
- ESD clamp element ESD robustness width scaling
- ESD on-resistance

8.6.8 ESD Power Clamp Trigger Condition versus Shunt Failure

For proper operation of the ESD power clamp, the trigger network will require to initiate prior to the overvoltage or overcurrent of the ESD "shunt" clamp element [18]. In the frequency domain, if the ESD network trigger does not respond to a specific ESD event, the trigger network will not respond effectively, and the ESD "shunt clamp" will discharge according to its native breakdown event. For a MOSFET "shunt" element, the element will undergo MOSFET drain-to-source snapback. For a bipolar transistor, the bipolar element will undergo collector-to-emitter breakdown.

8.6.9 ESD Clamp Element: Width Scaling

It is desirable to have the ESD results scale with the ESD clamp "shunt" element size. The ESD robustness will scale with the physical width given the following conditions:

- **Frequency tuning:** Proper frequency "tuning" of the trigger network (e.g., responsive to the ESD event) for MOSFET gate-driven networks or bipolar base-driven networks

- **Drive circuit:** Adequate current drive and current-drive distribution for bipolar base-driven networks

- **Layout symmetry:** Layout optimization of clamp element

- **Ballasting:** MOSFET drain ballast (or bipolar emitter ballast) adequate to provide uniformity

- **Power bus connectivity:** Electrical connection to power bus and ground rail well distributed in the ESD power clamp "clamp element" region of the circuit

8.6.10 ESD Clamp Element: On-Resistance

It is desirable to have the ESD clamp on-resistance which reduces with the size of the MOSFET or bipolar clamp element. The lower the ESD clamp on-resistance, the lower the total resistance through the alternative current loop. The lower the resistance in the ESD current loop, the lower the node voltage at the bond pad node. As the impedance of the power bus and the ESD clamp element is reduced, the allowed resistance for the ESD signal pin network can be higher and achieve the same signal pin ESD robustness. Hence, lowering the ESD power clamp resistance allows for a smaller ESD network at the signal pin (e.g., smaller network with lower capacitance).

The ESD clamp on-resistance will scale down with the clamp element device size, given that the element does not undergo current saturation effects, self-heating, or poor current distribution. Hence, if the ESD power clamp element is large enough and self-heating is kept to a minimum, the "on-resistance" will scale with the width scaling.

8.6.11 ESD Clamp Element: Safe Operating Area

The ESD "clamp element" must remain in the safe operating area (SOA) of the device to avoid failure of the ESD power clamp network. To avoid electrical failure of the ESD clamp element prior to the achieve ESD objective, the clamp element of the ESD power clamp must remain below a voltage absolute maximum ($V_{\text{ABS MAX}}$) and a current absolute maximum ($I_{\text{ABS MAX}}$) of the clamp element.

8.7 ESD POWER CLAMP ISSUES

ESD power clamps have some unique issues as a result of being placed within the power grid of a semiconductor chip. The issues will be briefly discussed, followed by examples in the future sections on how to address these issues.

8.7.1 Power-Up and Power-Down

ESD power clamps are to remain in an "off-state" when a semiconductor chip is in a power-up state, a power-down state, and a quiescent powered state [18]. The different solutions to avoid initiation of the power clamps during power-up and power-down ramping are as follows:

- **Frequency window:** Trigger networks do not respond to these frequencies.

- **Feedback networks:** Feedback networks are placed to avoid response to power up.

- **Enable/disable functions:** Logic can be integrated into the trigger network to "enable" or "disable" the ESD power clamp as desired.

8.7.2 False Triggering

ESD power clamps can be "false triggered" as a result of pulse events from signals, overcurrent, overvoltage, or "spikes" during test, burn-in, or other reliability stresses. The different solutions to avoid initiation of the power clamps during power-up and power-down are as follows:

- **Overcurrent protection:** Overcurrent protection can be integrated to avoid the ESD power clamp outside of its SOA.

- **Frequency window:** Trigger networks do not respond to these frequencies of "spikes."

- **Feedback networks:** Feedback networks are placed with hysteresis.

- **Enable/disable functions:** Logic can be integrated into the trigger network to "enable" or "disable" the ESD power clamp as desired.

8.7.3 Precharging

Precharging events can occur during ESD testing that can influence the ESD power clamp networks [18]. In the process of ESD testing, poor isolation of the test source from the device under test (DUT) can lead to a precharging phenomenon in the semiconductor chip. After an ESD pulse is applied, a low-level current bleeds from the HV source to the DUT without proper "switch" isolation. The solutions for the "precharging" solution are as follows:

- **ESD power clamp precharge "bleed" device:** A high impedance element can be placed in parallel to the ESD power clamp to allow the bleeding of charge from the V_{DD} to the V_{SS} power rail. The "bleed device" can be a resistor. This can be placed local to the device or nonlocal to the ESD power clamp.

- **ESD test system modification:** Modification of the ESD stress test system by providing proper isolation.

8.7.4 Postcharging

A postcharging event from ESD simulators is also present that can influence the ESD test results [18]. After the ESD event occurs, a low-level current "tail" exists in the simulators that continues to charge the signal pins or power pins. In the process of ESD testing, poor isolation of the test source from the DUT can lead to a postcharging phenomenon in the semiconductor chip. As in the precharging event, the postcharging events can lead to an anomalous ESD test result.

8.8 ESD POWER CLAMP DESIGN

In this section, examples of different circuit topologies will be shown to highlight some of previously discussed issues. Native power supply voltage and nonnative ESD power clamps will be discussed.

8.8.1 Native Power Supply RC-Triggered MOSFET ESD Power Clamp

Figure 8.18 shows an example of the most commonly used ESD power clamp in the semiconductor industry, the RC-triggered MOSFET ESD power clamp. The RC discriminator network discriminates between ESD events and spurious events or power-up and power-down if properly tuned. The RC trigger typically is "tuned" to respond to the ESD HBM and MM pulse events.

The inverter stages serve two purposes. Firstly, it allows for the tuning of the RC network without the loading of the first inverter gate capacitance influencing the RC "tuning." Secondly, it serves a drive stage for "driving" the ESD clamp element. In recent years, to

Frequency-dependent trigger network

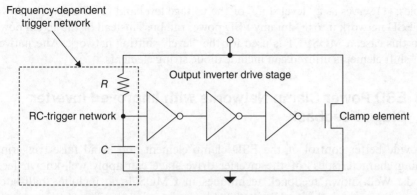

Figure 8.18   RC-triggered MOSFET ESD power clamp.

improve the responsiveness, the three inverter stages have been reduced to a single stage. The advantage of this is to improve the responsiveness. The disadvantage of the single inverter is the increase in the size of the single inverter stage and the lack of isolation between the RC discriminator tuning and the load of the inverter stage and output network. This network is also suitable for native voltage conditions. Given higher-voltage power domains, all elements in the circuitry must be voltage tolerant to that given power domain.

8.8.2 Nonnative Power Supply RC-Triggered MOSFET ESD Power Clamp

In many mixed-voltage or mixed-signal applications, different power clamps are required based on the voltage of the power domain. Figure 8.19 shows an RC-triggered MOSFET power clamp, where the second MOSFET is used to lower the voltage across all the elements in the lower element. In the design synthesis of this network, the "drop-down" device lowers the voltage across all elements in the ESD power clamp. Hence, it provides

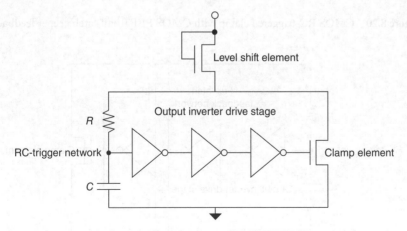

Figure 8.19   Series cascode RC-triggered MOSFET ESD power clamp.

two roles: (1) serves as a "level shift" of the voltage level and (2) converts the power bus of the ESD network into a "dummy ESD power rail bus" instead of the actual power rail bus. In this case, a MOSFET is used for the "level" shifting network. Alternative ESD "level" shift elements utilized can include diode string elements.

8.8.3 ESD Power Clamp Networks with Improved Inverter Stage Feedback

To provide better control of the ESD clamp element and avoid false triggering, the "latching characteristics" of the inverter drive stage can apply well-known feedback methods. Well-known feedback techniques in CMOS logic include "half-latch" or "full-latch" circuit concepts. Figure 8.20 shows an example of an ESD power clamp with a CMOS half-latch PMOS keeper element. This provides improved control of the MOSFET output gate, which can improve intolerance to false triggering or avoid low-level leakage of the output MOSFET.

The second method to improve the "latching characteristics" of the inverter drive stage can apply well-known "full-latch" circuit concepts. Figure 8.21 shows an example

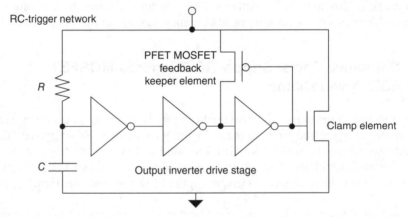

Figure 8.20 CMOS RC-triggered clamp with CMOS PFET half-latch keeper feedback.

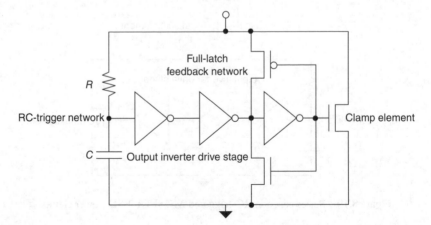

Figure 8.21 CMOS RC-triggered clamp with CMOS PFET full-latch keeper feedback.

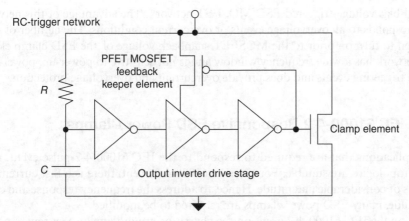

Figure 8.22   CMOS RC-triggered clamp with CMOS PFET cascade feedback.

of an ESD power clamp with a CMOS full-latch feedback network [18]. The integration of the full inverter for the feedback forms a "SRAM-like" latch between the ESD power clamp last inverter and the feedback inverter. As with the "half-latch" feedback, this provides improved control of the MOSFET output gate, which can improve intolerance to false triggering or avoid low-level leakage of the output MOSFET.

Other techniques for improving the control of the ESD power clamp from false triggering can be applied. As the feedback is brought to the earlier stages, the size of the feedback elements can be reduced. The third method is placement of a PMOS device above the inverters. Figure 8.22 shows an example of an ESD power clamp with a PMOS element within the logic [18].

8.8.4 Forward-Bias Triggered ESD Power Clamps

In some applications, the presence of a frequency-triggered network is undesirable. RC-triggered ESD MOSFET networks provide ESD protection for frequency dependent applications [18]. For example, given the frequency response of the system, such as a cell phone, is predefined, it may be not be advisable to place another frequency-dependent circuit in a small system (e.g., altering the frequency response of the poles and zeros in the frequency domain). As a result, some circuit design teams desire voltage-triggered networks for RF CMOS instead of frequency-triggered networks. Figure 8.23 is an example of a

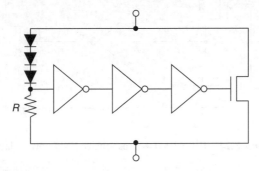

Figure 8.23   Forward-bias voltage-triggered ESD power clamps.

forward-bias voltage-triggered ESD MOSFET network. The advantage of this network is that it responds to all overvoltage events or overcurrent conditions. The number of diodes is chosen to turn on prior to the MOSFET snapback voltage of the ESD clamp element. This network has a wide frequency window and is not sensitive to power-up, power-down, or false triggering events and does provide overcurrent and overvoltage protection.

8.8.5 IEC 61000-4-2 Responsive ESD Power Clamps

For applications that are required to respond to the IEC 61000-4-2 pulse event, not all circuit topologies are suitable. For the IEC 61000-4-2 event, there is a fast current pulse which is of considerable magnitude. Hence, to address the frequency response and current magnitude, many ESD power clamps are required to be modified.

During the IEC 61000-4-2 event on the chassis or ground line of a system, a negative pulse occurs on the V_{SS} power rail or substrate. This can initiate the RC-triggered network from the negative pulse event. But the elements in the RC discriminator must be responsive, or circuit failure can occur. The resistor and capacitor element choices must be responsive. Resistors, such as polysilicon resistors, may be slow to respond to fast events.

Figure 8.24 is an example of an IEC 61000-4-2 event responsive ESD MOSFET network. The advantage of this network is that the p-channel MOSFET is more responsive than a polysilicon resistor element. Additionally, so that the inverter drive network is more responsive, only a single inverter stage is implemented.

8.8.6 Precharging and Postcharging Insensitive ESD Power Clamps

ESD test systems or residual charge can influence the "state" of an RC-triggered MOSFET clamp before or after ESD stress. With charge on the V_{DD} power rail, the voltage state of the RC-triggered MOSFET can be precharged and close the MOSFET snapback voltage of the ESD clamp device. On the first discovery of this issue, it was noted by *R*. Ashton

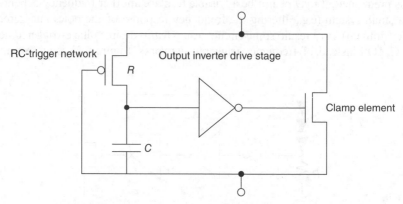

Figure 8.24   IEC 61000-4-2 responsive ESD power clamp.

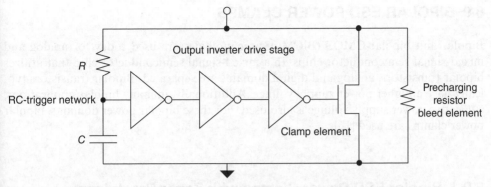

Figure 8.25   Precharging and postcharging insensitive ESD power clamp.

that products, with RC-triggered power clamps, that were inherently "leaky" had better ESD results than products whose V_{DD} leakage was low. It was from this discovery that Ashton discovered the issue of ESD test system leading to residual charge on the power grid of the semiconductor chip, influencing the prestate of the ESD power clamp. It was noted that the charge on V_{DD} power rail leads to the MOSFET snapback of the output device prior to initiation of the RC discriminator response.

Figure 8.25 is an example of an ESD power clamp network with a "bleed" element to provide discharging of the ESD precharging event or a postcharging event. Placing a high impedance element that bleeds the charge off of the power rail can avoid the ESD test system-induced operation failure of the ESD power clamp element.

8.8.7 ESD Power Clamp Design Synthesis and Return Diode

ESD power clamps require bidirectionality to allow current flow in a first direction from analog or digital V_{DD} to V_{ss} and a second direction from V_{ss} to analog or digital V_{DD}. As a result, a "return diode" is added to provide current flow in the reverse direction. Figure 8.26 is an example of a "return diode" ESD network that uses an LV CMOS diode in parallel with an RC-triggered ESD power clamp.

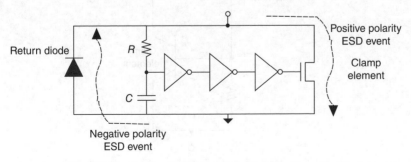

Figure 8.26   Analog ESD power clamp and return diode.

8.9 BIPOLAR ESD POWER CLAMPS

Bipolar and bipolar-CMOS (BiCMOS) technologies are used today for analog and mixed-signal semiconductor chips. In a mixed-signal semiconductor chip that utilizes bipolar transistors, analog and digital domains are separated. Bipolar transistors typically have a higher power supply voltage. Additionally, in many bipolar applications, a negative power supply voltage is also used. For these bipolar power domains, bipolar power clamps are used [16].

8.9.1 Bipolar ESD Power Clamps with Zener Breakdown Trigger Element

Figure 8.27 is an example of a bipolar ESD power clamp [16]. In this ESD bipolar power clamp, a single transistor is placed between the two power supplies in a collector-to-emitter configuration. The transistor is to be used to discharge the ESD current from the V_{CC} power rail to the V_{SS} ground rail. The trigger element is a Zener diode which undergoes electrical breakdown. When the voltage across the trigger element reaches the breakdown voltage of the Zener diode, the current flows through the Zener diode and into the base of the bipolar transistor. This base-driven network responds to overvoltage conditions in the semiconductor chip. Since it is a voltage-triggered network, it has a wide frequency window of operation; the frequency response is limited to the frequency response of the Zener diode and its bipolar transistor.

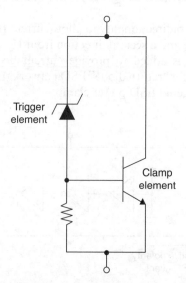

Figure 8.27 Reverse breakdown Zener-triggered bipolar ESD power clamps.

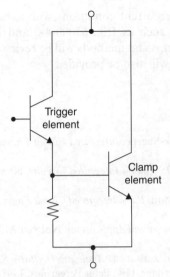

Figure 8.28 Bipolar ESD power clamps with BV_{CEO} breakdown trigger element.

8.9.2 Bipolar ESD Power Clamps with Bipolar Transistor BV_{CEO} Breakdown Trigger Element

Figure 8.28 is an example of a BV_{CEO} voltage-triggered bipolar ESD power clamp. In this ESD bipolar power clamp, a single transistor is placed between the two power supplies in a collector-to-emitter configuration [16]. A second transistor is used as the trigger element and is also placed in a common-emitter (C-E) configuration. The clamp transistor element is a high-breakdown (HB) transistor and is to be used to discharge the ESD current from the V_{CC} power rail to the V_{SS} ground rail. The trigger element is a low breakdown (LB) voltage (e.g., BV_{CEO}) npn transistor which undergoes electrical breakdown. When the voltage across the trigger element reaches the BV_{CEO} breakdown voltage, the current flows through the trigger element and into the base of the bipolar transistor. This base-driven network responds to overvoltage conditions in the semiconductor chip. Since it is a voltage-triggered network, it has a wide frequency window of operation; the frequency response is limited to the frequency response of the two transistors.

8.10 CLOSING COMMENTS AND SUMMARY

This chapter focused on analog and ESD integration. It also discussed analog signal pin input circuitry to ESD power clamps. A more in-depth look at ESD power clamp issues and solutions was shown.

Chapter 9 discusses system-level issues associated with EOS in chips, printed circuit boards (PCBs), and systems. EOS protection device classifications, symbols, and types for both overvoltage and overcurrent conditions will be highlighted. System-level and system-like testing methods, such as IEC 61000-4-2 and IEC 61000-4-5, and human metal model (HMM) waveforms and methods will be reviewed. Examples of PCB design for digital-to-analog systems will also be provided.

REFERENCES

1. A.B. Glasser and G.E. Subak-Sharpe. *Integrated Circuit Engineering*. Reading, MA: Addison-Wesley, 1977.
2. A. Grebene. *Bipolar and MOS Analog Integrated Circuits*. New York: John Wiley & Sons, Inc., 1984.
3. D.J. Hamilton and W.G. Howard. *Basic Integrated Circuit Engineering*. New York: McGraw-Hill, 1975.
4. A. Alvarez. *BiCMOS Technology and Applications*. Norwell, MA: Kluwer Academic Publishers, 1989.
5. R.S. Soin, F. Maloberti, and J. Franca. *Analogue-Digital ASICs, Circuit Techniques, Design Tools, and Applications*. Stevenage, UK: Peter Peregrinus, 1991.
6. P.R. Gray and R.G. Meyer. *Analysis and Design of Analog Integrated Circuits*. 3rd Edition. New York: John Wiley & Sons, Inc., 1993.
7. F. Maloberti. Layout of analog and mixed analog-digital circuits. In: J. Franca and Y. Tsividis (Eds). *Design of Analog-Digital VLSI Circuits for Telecommunication and Signal Processing*. Upper Saddle River, NJ: Prentice Hall, 1994.
8. D.A. Johns and K. Martin. *Analog Integrated Circuit Design*. New York: John Wiley & Sons, Inc., 1997.
9. R. Geiger, P. Allen, and N. Strader. *VLSI: Design Techniques for Analog and Digital Circuits*. New York: McGraw-Hill, 1990.
10. A. Hastings. *The Art of Analog Layout*. Upper Saddle River, NJ: Prentice Hall, 2006.
11. V. Vashchenko and A. Shibkov. *ESD Design for Analog Circuits*. New York: Springer, 2010.
12. S. Voldman. *Electrical Overstress (EOS): Devices, Circuits, and Systems*. Chichester, UK: John Wiley & Sons, Ltd, 2013.
13. S. Voldman. *ESD Basics: From Semiconductor Manufacturing to Product Use*. Chichester, UK: John Wiley & Sons, Ltd, 2012.
14. S. Voldman. *ESD: Physics and Devices*. Chichester, UK: John Wiley & Sons, Ltd, 2004.
15. S. Voldman. *ESD: Circuits and Devices*. Chichester, UK: John Wiley & Sons, Ltd, 2005.
16. S. Voldman. *ESD: RF Circuits and Technology*. Chichester, UK: John Wiley & Sons, Ltd, 2006.
17. S. Voldman. *ESD: Failure Mechanisms and Models*. Chichester, UK: John Wiley & Sons, Ltd, 2009.
18. S. Voldman. *ESD: Design and Synthesis*. Chichester, UK: John Wiley & Sons, Ltd, 2011.

9 System-Level EOS Issues for Analog Design

9.1 EOS PROTECTION DEVICES

Electrical overstress (EOS) protection devices are supported by a large variety of analog technologies [1–24]. Although material and operation may differ between the EOS protection devices, their electrical characteristics can be classified into a few fundamental groups. Figure 9.1 shows the classification of EOS devices.

EOS protection networks can be identified as a voltage suppression device or as a current-limiting device. The voltage suppression device limits the voltage observed on the signal pins or power rails of a component, preventing electrical overvoltage (EOV). The current-limiting device prevents a high current from reaching sensitive nodes, avoiding electrical overcurrent (EOC).

9.1.1 EOS Protection Device: Voltage Suppression Devices

Voltage suppression devices can also be subdivided into two major classifications [1, 19]. Figure 9.2 illustrates the examples of voltage suppression categories. Voltage suppression devices can be segmented into devices that remain with a positive differential resistance and those that undergo a negative resistance region. For positive differential resistance, these devices can be referred to as "voltage clamp" devices where dI/dV remains positive for all states; for the second group, there exists a region where dI/dV is negative. The first group can be classified as "voltage clamp devices," whereas the second group can be referred to as an "S-type I–V characteristic device" or as a "snapback device." In the classification of voltage suppression devices, the second classification can be associated with the directionality; a voltage suppression device can be "unidirectional" or "bidirectional" (Figure 9.3).

ESD: Analog Circuits and Design, First Edition. Steven H. Voldman.
© 2015 John Wiley & Sons, Ltd. Published 2015 by John Wiley & Sons, Ltd.

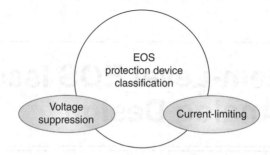

Figure 9.1   EOS protection device classifications.

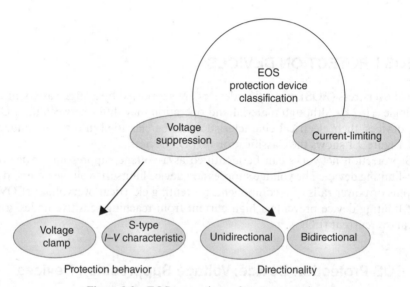

Figure 9.2   EOS protection voltage suppression.

9.1.2 EOS Protection Device: Current-Limiting Devices

Current-limiting devices can be used in a series configuration for EOS protection. EOS current-limiting devices can be as follows [1, 19–24]:

- Resistors
- Resetting fuses
- Nonresetting fuses
- Electronic fuse (eFUSE)
- Positive temperature coefficient (PTC) devices
- Circuit breakers

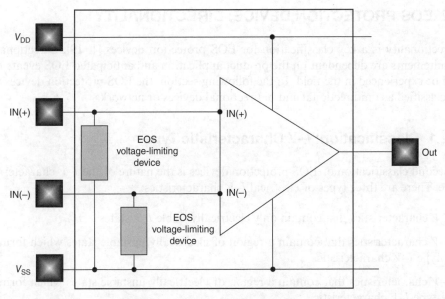

Figure 9.3   Voltage-limiting EOV devices.

The choice of the current-limiting EOS protection device is a function of the cost, size, rated current, time response, I^2t value, rated voltage, voltage drops, and application requirements.

Figure 9.4 illustrates an example of an EOC protection device solution integrated with a differential circuit.

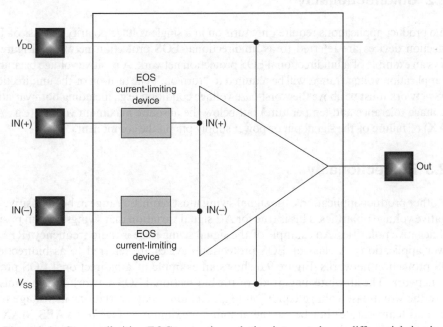

Figure 9.4   Current-limiting EOC protection solution integrated on a differential circuit.

9.2 EOS PROTECTION DEVICE: DIRECTIONALITY

Directionality is a key classification for EOS protection devices [1, 19]. Directionality requirements are dependent on the product application and anticipated EOS events that will be experienced in the field. In the following section, the EOS protection devices will be classified as unidirectional and bidirectional devices or networks.

9.2.1 Classification: *I–V* Characteristic Type

A second classification of EOS protection devices is the nature of the *I–V* characteristic type. There are three types of electrical *I–V* characteristics:

- *I–V* characteristics that contain only electrically stable *I–V* states

- *I–V* characteristics that contain a region of electrically unstable states which form an S-type *I–V* characteristic

- *I–V* characteristics that contain a region of electrically unstable states which form an N-type *I–V* characteristic

The first type has only positive resistance in all *I–V* states. For the second and third *I–V* types, there are regions of negative resistance, which are electrically unstable conditions [1, 19]. For EOS networks, the focus will only be on the first two *I–V* characteristic types.

9.2.2 Unidirectionality

Some product applications require only turn-on in a single voltage polarity. A class of EOS protection devices are referred to as unidirectional EOS protection networks. Figure 9.5 shows an example of a unidirectional EOS protection network. A positive voltage bias, above the application voltage range, will be required to "turn on." The turn-on of the unidirectional EOS network must be above the worst-case voltage condition (e.g., including both variations in voltage tolerance and temperature) but below the absolute maximum voltage (e.g., ABS MAX) of failure of the signal pin or power supply pin on the component.

9.2.3 Bidirectionality

For other product applications, the signal swing must require a range in both positive and negative voltage polarities. This is true for AC signal variation that swings both in positive and negative polarities. An example of the signal swing issue is radio frequency (RF) and power applications. A class of EOS protection devices are referred to as bidirectional EOS protection networks. Figure 9.6 shows an example of a bidirectional EOS protection network. The absolute magnitude of the bidirectional EOS network turn-on must be above the worst-case voltage condition (e.g., including both variations in voltage tolerance and temperature) but below the absolute maximum voltage (e.g., ABS MAX) of failure of the signal pin or power supply pin on the component for both polarities.

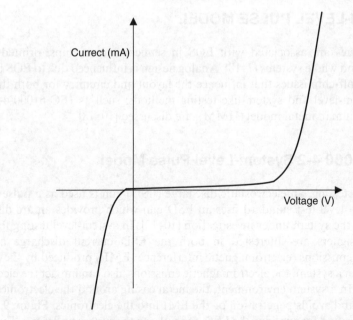

Figure 9.5   Unidirectional EOS protection device $I–V$ characteristic.

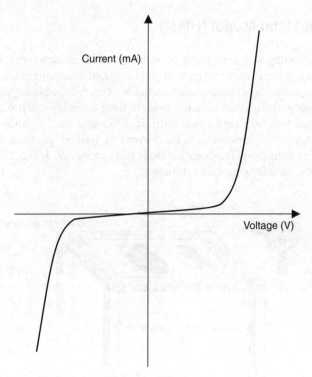

Figure 9.6   Bidirectional EOS protection device.

9.3 SYSTEM-LEVEL PULSE MODEL

System-level issues are associated with EOS in semiconductor chips, printed circuit boards (PCB), and whole systems [7–18]. Analog design is influenced due to EOS because of on-chip and off-chip issues that influence the layout and circuitry for both the chips and PCB. System-level and system-like testing methods, such as IEC 61000-4-2, IEC 61000-4-5, and human metal model (HMM), are discussed [10–18].

9.3.1 IEC 61000-4-2 System-Level Pulse Model

For system-level testing, an electrostatic discharge (ESD) gun is used as a pulse source. This IEC system-level test standard uses an ESD gun which provides an arc discharge from the gun to the system under investigation [10–13]. In system-level testing for ESD, system-level designers are interested in both the ESD current discharge and the electromagnetic emissions (electromagnetic interference (EMI)) produced by the arc discharge process. In a system, the electromagnetic emissions also can impact the electronics or components. In a system environment, the metal casing around the electronics forms a Faraday cage and avoids penetration of the EMI into the electronics. Figure 9.7 is the IEC test configuration for applying the ESD gun pulse to the system under test. Figure 9.8 shows the IEC 61000-4-2 waveform for the IEC test [10–13].

9.3.2 Human Metal Model (HMM)

In the past, ESD testing was performed on semiconductor components. As stated in the last section, today, there is more interest in the testing of components in powered states and in electrical systems. System manufacturers have begun requiring system-level testing to be done on semiconductor components, prior to final assembly and product acceptance. These system-level tests are performed with an ESD gun and without direct contact; these air discharge events produce an ESD event as well as generate EMI. In a true system, the system itself provides shielding from EMI emissions. Hence, an ESD test is of interest which has the following characteristics:

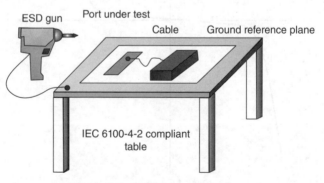

Figure 9.7 IEC test configuration.

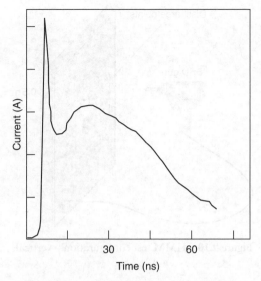

Figure 9.8   IEC 61000-4-2 current waveform.

- An IEC 61000-4-2 current waveform.

- No air discharge (contact discharge).

- Semiconductor component is powered during ESD testing.

- Only addresses pins and ports exposed to the external system.

The HMM addresses these characteristics [15–18]. The HMM event is a recent ESD model which has increased interest as a result of cell phone and small components with exposed ports, where field failures were evident. The HMM uses an "IEC-like" pulse waveform. The discharge from the source and the device under test (DUT) is a direct contact to avoid EMI spurious signals. The test is performed when the system is powered, and only the external ports that are exposed to the outside world are of interest. Figure 9.9 shows the test system horizontal configuration, where the source is an ESD gun. Figure 9.10 shows the test system vertical configuration, where the source is an ESD gun.

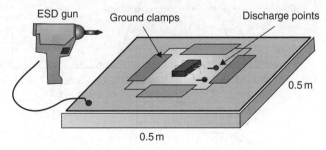

Figure 9.9   HMM test configuration—horizontal.

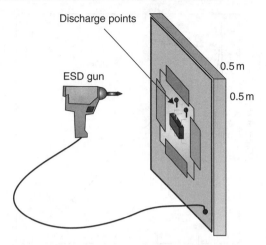

Figure 9.10 HMM test configuration—vertical.

For the HMM pulse, the source does not have to be an ESD gun. The method applies an IEC pulse to the DUT without any air discharge. Using a current source, variations in the ESD gun waveform and pulse variation are removed.

9.3.3 IEC 61000-4-5 Surge Test

For evaluation of EOS, standards have been developed for "surge testing" [14]. A standard that is being utilized today for evaluation of components or populated PCB is the IEC 61000-4-5 Electromagnetic Compatibility (EMC)—Part 4-5: Testing and Measurement Techniques—Surge Immunity Test standard [14].

Figure 9.11 shows the IEC 61000-4-5 surge test pulse open-circuit waveform. This pulse waveform is nonoscillatory for an open circuit.

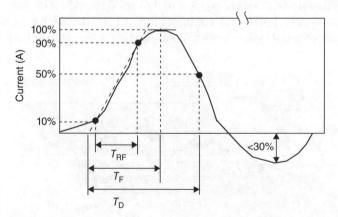

Figure 9.11 IEC 61000-4-5 surge test pulse open-circuit waveform.

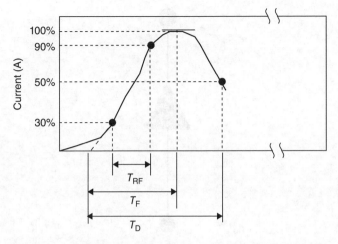

Figure 9.12   IEC 61000-4-5 surge test pulse short-circuit waveform.

Figure 9.12 shows the IEC 61000-4-5 surge test pulse short-circuit waveform. This pulse waveform undergoes a negative transition after the first peak.

9.4 EOS TRANSIENT VOLTAGE SUPPRESSION (TVS)

A major concern in EOS is associated with transient voltage conditions that occur in the assembly and production. Transient voltage conditions also occur in test as well as burn-in. During burn-in, products are stressed above the power supply voltage of the application (e.g., $1.5 \times V_{DD}$). Additionally, noise and spikes occur in burn-in chambers. To suppress voltage transients, different diodes are used [1]:

• Silicon p–n junction diodes

• Schottky diodes

9.4.1 EOS Diodes

Diodes are a commonly used unidirectional EOS protection device. Diodes provide a forward conduction state and reverse blocking state. For EOS, single-component diodes are mounted on PCB by soldering in the leads through vias or surface mount. Diodes are commonly used within components to provide ESD protection for internal circuitry [1]. The symbol for a diode element is shown in Figure 9.13.

Electrical characteristics of a diode structure include forward turn-on, the on-resistance, and reverse breakdown voltage. At high-current conditions, in the forward conduction state, thermal breakdown occurs as well as thermal failure of the diode structure. If the EOS stress current exceeds the thermal breakdown of the diode structure, destructive failure of the element can lead to system degradation or an EOS failure.

Figure 9.13   Diode symbol.

Diodes are unidirectional-type EOS structure but can be utilized in a forward or reverse breakdown mode of operation for a voltage-limiting EOS solution [1]. The power-to-failure of a diode structure is higher in the forward mode of operation compared to its reverse breakdown mode of operation. In a forward conduction mode, the power distributes through all sections of the element (e.g., anode, cathode, and metallurgical junction). In the reverse conduction mode, power is contained only in the metallurgical junction region.

9.4.2 EOS Schottky Diodes

Schottky diodes are also commonly used unidirectional EOS protection device [1]. Schottky diodes have a forward conduction state and a reverse blocking state. Schottky diodes have a lower forward turn-on (e.g., 0.35 V) compared to standard silicon p–n junction (e.g., 0.7 V).

For EOS, single-component Schottky diodes are mounted on PCB by soldering in the leads through vias or surface mount. Schottky diodes are not as commonly used within components to provide ESD protection due to lack of availability. Schottky diodes are unidirectional-type EOS structure but can be utilized in a forward or reverse breakdown mode of operation for a voltage-limiting EOS solution.

9.4.3 EOS Zener Diodes

Zener diodes are used as a unidirectional EOS protection device [1]. Zener diodes are typically used as a voltage clamping EOS protection device, and typically used in the breakdown state. Zener diodes have a well-defined voltage breakdown value.

Zener diodes are used for ESD protection for high-voltage and power applications. Zener diodes are not used for ESD protection for low-voltage CMOS applications. For EOS, single-component Zener diodes are mounted on PCB by soldering in the leads through vias or surface mount.

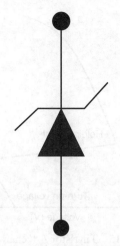

Figure 9.14   Zener diode.

Zener diodes are unidirectional-type EOS structure but can be utilized primarily in reverse breakdown mode of operation for a voltage-limiting EOS solution (Figure 9.14).

9.4.4 EOS Thyristor Surge Protection

Thyristors can be used as an EOS due to many of its electrical response and current-carrying capability [1, 19]. A thyristor is also known as a silicon-controlled rectifier (SCR) or pnpn device [1]. The thyristor has many advantages as an EOS device. Thyristors have a forward blocking state allowing for them to remain in the "off-state" during normal operating conditions. Thyristors can operate in a low-voltage high-current state with a low on-resistance. Thyristors can be unidirectional or bidirectional. Thyristors can be designed so they operate as a symmetrical bidirectional device. Thyristors also have a fast turn-on time to allow responsiveness to fast current transients or surges.

To serve as an EOS protection device, thyristors can be used for protection against surges; hence, they are referred to as a thyristor surge protection device (TSPD).

Thyristors are a four-region pnpn device. A pnpn device can be formed as a pnp transistor and a npn transistor cross coupled. These resistor elements can be intrinsic or extrinsic elements.

The cross coupling between the pnp regions and the npn regions allows for feedback between the two transistors. The feedback between the two elements leads to an unstable negative resistance region, forming an S-type characteristic. The thyristor is also called an SCR (Figure 9.15).

9.4.5 EOS Metal-Oxide Varistors (MOV)

An EOS protection device used for high voltages is the varistor [1, 19]. A varistor is also known as a voltage-dependent resistor (VDR). A varistor is a portmanteau—combining the word for variable and resistor. In reality, the varistor element behaves like a diode,

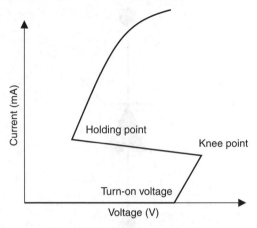

Figure 9.15   Thyristor $I\text{--}V$ characteristic.

forming a nonlinear current–voltage ($I\text{--}V$ characteristic). The element has the characteristics of being bidirectional voltage clamp (note: it does not form an S-type $I\text{--}V$ characteristic nor undergo a negative resistance regime). At low voltages, the device has a high series resistance, serving as an "off-state." At higher voltages, the device "turns on" and has a low resistance.

The metal-oxide varistor (MOV) device is the most common varistor composition. Figure 9.16 is the circuit schematic symbol for an MOV device. Zinc oxide, combined with other metal oxides, is integrated between two metal electrodes. Other metal oxides integrated include bismuth, cobalt, and manganese. The operation of the MOV device is based on conduction through ZnO grains; current flows "diode-like" through the grain structures creating a low-current flow at low voltages. At higher voltages, the current flow is dominated by a combination of thermionic emissions and tunneling. This diode-like behavior that forms the diode-like characteristic provides the high-resistance/low-voltage state and the low-resistance/high-voltage state. From the $I\text{--}V$ characteristic, a "clamping voltage" can be defined (e.g., analogous to a diode turn-on voltage). The characteristics that influence the on-resistance and the turn-on voltage are a function of the ZnO grain structure, the film thickness, the physical size, and the other metal oxides integrated into the structure.

The advantage of the MOV structure is it has a high trigger voltage, making it suitable for EOS protection in power electronics (e.g., 120–700 V applications). The disadvantage

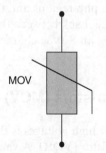

Figure 9.16   MOV symbol.

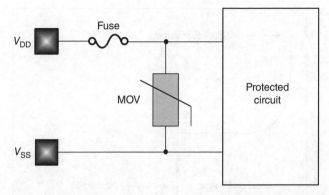

Figure 9.17   Power supply protection with MOV EOS device.

of these elements is that it has high capacitance and high on-resistance. A second disadvantage is that the high trigger voltage does not make it suitable for advanced high-speed or low-voltage electronics. A third disadvantage is the variability of the device response (e.g., on-resistance and clamping voltage) in the MOV device characteristics. Key device parameters of varistor are the energy rating, operating voltage, response time, maximum current, and breakdown voltages.

Figure 9.17 is an example of usage of an MOV for power supply protection. The MOV element can be placed between the power supply and ground to avoid EOS to the V_{DD} power supply.

9.4.6 EOS Gas Discharge Tubes (GDT)

Gas discharge tube (GDT) devices can be used to avoid EOS in systems [1, 19]. GDT are bidirectional, allowing for protection for both positive and negative EOS events. GDT elements are suitable from surge protection to high-voltage electronic "crowbars" to lightning. GDT devices are not suitable for EOS protection of low-voltage electronics due to the high trigger voltages (unless used as a first stage followed by other low-voltage secondary EOS solutions). Figure 9.18 shows the symbol for a GDT.

Gas-filled tubes (GDT) utilize electrical discharge in gases. An applied voltage initiates the device by ionizing the electrical gas, followed by electrical glow discharge, and an electrical arc. With creation of an electrical arc, the GDT device becomes a low-resistance

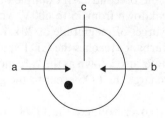

Figure 9.18   GDT device symbol.

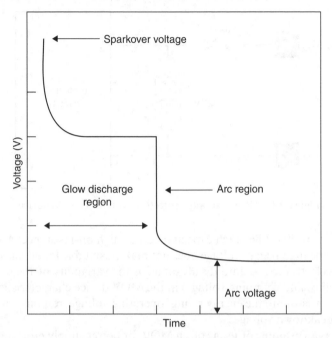

Figure 9.19   Switching characteristic of a GDT device—ramp impulse voltage versus time.

shunt for EOS protection. These gas-filled tubes can contain hydrogen, deuterium, and noble gases (e.g., helium, neon, argon, krypton, and xenon). GDT devices can vary their electrical characteristics by choices of the gas type, pressure, electrode design, and spacings.

Figure 9.19 shows an example of a ramp impulse voltage versus time for a GDT device. GDT devices undergo three states: (1) electrical breakdown, (2) glow discharge, and (3) electrical arc. The electrical breakdown is a high-voltage low-current state prior to triggering of the GDT device. A glow discharge region forms a second state which incorporates a low-current high-voltage state. Lastly, after full ionization of the gas, a low-voltage high-current state occurs with a low "on-resistance."

GDT devices have high trigger voltages suitable for LDMOS power electronic applications to HV-LDMOS (e.g., 120 V) and UHV-LDMOS applications (e.g., 600–700 V). These devices are used in a number of high-voltage switch devices, such as ignitrons, krytrons, and thyratrons. One of the disadvantages of the GDT devices is the slow turn-on times typically in the microseconds. An example of some of the electrical characteristics can exhibit DC breakdown from 75 to 600 V, with a single surge response of 40 kA in 10–20 µs or multiple surges of magnitude of 20 kA.

In time, when a voltage disturbance reaches the GDT sparkover voltage, the GDT will switch into a low-impedance state, also known as the "arc mode." With this low-resistance state, the GDT device will discharge the EOS event to the ground, avoiding EOS failure of the component.

At voltages below the sparkover voltage, the GDT remains in a high-impedance off-state. As the voltage increases, the GDT enters the "glow voltage region"; this glow region is where ionization starts to occur within the GDT. As the current increases, avalanche multiplication occurs, leading to further ionization of the gas within the GDT. This is followed by avalanche breakdown and the low-impedance state. The voltage condition developed across the GDT during this state is called the "arc voltage."

The transition time between the glow and arc region is dependent on the impulse current, electrode shape (e.g., electrode curvature), electrode spacing, gas composition, gas pressure, and emission coatings.

9.5 EOS CURRENT SUPPRESSION DEVICES

To prevent EOS associated with EOC, current suppression devices are mounted on PCB or integrated into systems. The choice of EOS device to use in an application is dependent on the electrical characteristics, cost, and size. The electrical characteristic that is of most importance with current-limiting devices is the resistance when the EOS device is operational.

Types of current-limiting devices used for EOS are as follows [19–24]:

- PTC devices

- Conductive polymers

- Fuses

- eFUSE

- Circuit breakers

9.5.1 EOS PTC Device

A class of current-limiting protection devices include elements whose resistance increases with increasing temperature. Devices that have this characteristic are valuable as limiting the current within a device or circuit. Current-limiting devices that have a PTC are used in both EOS and ESD protections. In the case of EOS events, these PTC elements may be separate components on the PCB. In the case of ESD, these may be an integrated circuit component that utilizes the natural material properties of the material.

Resistance is defined according to the linear relationship

$$R(T) = \mathrm{Ro}(1 + \alpha T)$$

where $R(T)$ is the dynamic resistance at the temperature T, Ro is the initial resistance, and α is the temperature coefficient of resistance (TCR). If the TCR is positive, the resistance will increase with increased temperature. We can relate temperature to the resistance change as

$$T = \frac{R(T) - \mathrm{Ro}}{\alpha \mathrm{Ro}} = \frac{1}{\alpha}\left(\frac{\Delta R}{\mathrm{Ro}}\right)$$

The heat flux is equal to the input power from conservation of energy. Substituting in power for the heat flux, the total differential resistance can be expressed as

$$\frac{dR}{Ro} = \alpha(Pd\theta + \theta dP)$$

Since the impedance is constant, then after integration, the expression from initial to final resistance can be expressed as

$$\frac{\Delta R}{R} = \frac{Rf - Ro}{Ro} = \alpha\theta_{TH}P$$

Hence, the normalized resistance change is proportional to the product of the thermal impedance, power, and TCR. Combining Joule heating (e.g., $P = I^2R$) and resistance ($R = \rho L/A$) and substituting it for the current density, J, then the expression of normalized differential resistance can be expressed as [2]

$$\frac{\Delta R}{R} = \alpha\theta_{TH} \, J^2\rho LA$$

From the normalized differential resistance, we can solve for the thermal impedance as

$$\theta_{TH} = \frac{1}{J^2} \frac{[\Delta R/R]}{\alpha\rho LA}$$

From this formulation, the thermal impedance can be extracted from the resistance change and the current density.

For ESD protection, current-limiting elements can utilize silicon and wire interconnects as resistance elements. Series resistors can be placed in series between bond pads and sensitive circuits. Series resistors can also serve as ballasting elements. For wire interconnect resistors, such as tungsten (W), tungsten wiring can serve as a high melting temperature element, whose resistance increases with higher temperature.

For EOS, single components on a PCB as series elements using PTC elements such as conductive polymers can be used.

9.5.2 EOS Conductive Polymer

Voltage suppression devices are needed in electronic systems to prevent damage to electrical components from EOS. One solution for voltage suppression is the conductive polymer surge protection device utilizing conductive polymers.

A low-capacitance, polymer voltage suppression (PVS) device can have potential advantages in future technologies. With the trend toward smaller semiconductor chip geometries, higher functional frequencies, and the escalating number of signal lines, system-level EOS issues on products have increased. The advantages of PVS are as follows: it is of low capacitance, can be integrated into space-efficient packages, and does not utilize significant space on PCB. In PVS components, the low dielectric constant of the polymer film provides inherently low capacitance in the femtofarad (fF) magnitude; this opens opportunities as an EOS solution for RF and high-speed digital applications.

Single-unit PVS surface-mount device consists of a polymer film laminated between electrodes. Since PVS devices are bidirectional, a single device soldered on a PCB or component signal line shunts both positive and negative voltage surges from signal line through the PVS to ground.

For example, the PVS film used consists of an epoxy polymer, containing uniformly dispersed conductive and nonconductive particles, laminated between electrodes. The laminate is transformed into a device using a process similar to that used for manufacturing PCB. The PVS device overall thickness is in the range less than 10 mils. The PVS material does not require a substrate, and consequently, the trigger voltage of the device is set by device dimensions and the polymer formula. From the experimental results, the average capacitance of this surface-mount device is approximately 105 fF. With this device, the trigger voltage is 200 V [19]. Hence, this component is suitable for reduction of high-voltage EOS events, but is not suitable to provide ESD protection for low-voltage components. Additional difficulties with these components are the reliability to survive many cycles and poor on-state resistance.

9.5.3 EOS Fuses

Electrical fuses are still used today to solve problems of EOC in electronic systems from industrial, commercial, residential, and semiconductor-based electronics.

Two major classifications of fuses exist [23, 24]:

- Self-resetting fuses

- Nonresettable fuses

Self-resetting fuses are used in applications where fuse replacement is not possible; this can include aerospace, military, or nuclear equipment. Self-resetting fuses can utilize conductive polymers or thermoplastic elements such as a polymeric positive temperature coefficient (PPTC) thermistor element. The value of a self-resetting fuse is that the element restores to its original state after the EOC event has cleared.

Nonresettable fuses are more commonly used where the fuse element is destructive and needs to be replaced after initiation. Nonresettable fuses typically consist of a metallic element which melts during EOC events.

Electrical characteristic parameters of fuse elements are [1]:

- Rated current I_N

- Speed

- I^2t value

- Breaking capacity

- Rated voltage

- Voltage drop

- Temperature derating

9.5.3.1 Rated Current I_N
This electrical parameter is the current rating for the maximum current for the fuse during normal operation. This current magnitude is the allowed current magnitude without interruption to the component or system.

9.5.3.2 Speed
This electrical parameter is operation time of the fuse structure. Fuses are classified as slow blow, fast blow, ultrafast blow, and time delay. For semiconductor and low-voltage applications, it is important to use fast blow or ultrafast blow fuse elements. With fuses, the fuse speed is a function of the current magnitude.

9.5.3.3 I^2t Value
This parameter is associated with the energy utilized by the fuse to clear the EOC event. There are two parameters of interest: the melting I^2t parameter and the clearing I^2t. The melting I^2t parameter is associated with the amount of energy to melt the fuse element. The clearing I^2t is the amount of energy that is allowed through the fuse to clear the EOC fault event.

9.5.3.4 Breaking Capacity
The breaking capacity is associated with the maximum current level that can be addressed by the fuse element.

9.5.3.5 Rated Voltage
The rated voltage is the maximum voltage that a fuse can sustain an open circuit. Since the electrode spacings in the fuse have a finite distance, if the voltage is significantly high, the failed fuse can still only handle a certain voltage magnitude. The rated voltage must be greater than the maximum applied voltage in the circuit.

9.5.3.6 Voltage Drop
The fuse element has a series resistance. This series resistance introduces a voltage drop in the circuit. This is a larger issue with low-voltage semiconductor-based systems.

9.5.3.7 Temperature Derating
The fuse element is typically a metal element, whose resistance is a function of the ambient temperature. The ambient temperature (with no current applied) can influence

the electrical characteristics and change the allowed current level. At low temperatures, the current can be increased, and at high temperatures, the fuse must be derated for the temperature change.

Fuse standards exist internationally and in North America. The international fuse standard is IEC 60269; this is segmented into three sections—industrial and commercial, residential, and semiconductor protection [19]. The second standard is UL 248 [20], which is harmonized with the Canadian standard C22.2 No. 248.

9.5.4 EOS eFUSEs

Overcurrent to external loads can be prevented by using eFUSE [1]. eFUSEs are networks which are programmable for many types of functions. For EOS, current can be limited to preset values to prevent overcurrent. eFUSE networks can contain different functions as follows:

- Thermal shutdown control circuitry

- Voltage-limiting control circuitry

- Current-limiting control circuitry

Figure 9.20 is an example of an eFUSE network with control circuit functional blocks for avoiding EOV, EOC, and thermal runaway.

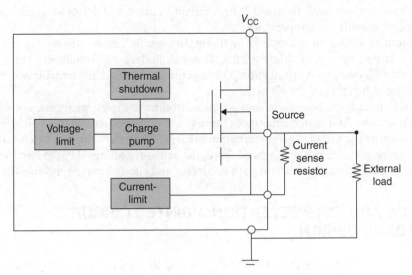

Figure 9.20   eFUSE EOS protection network.

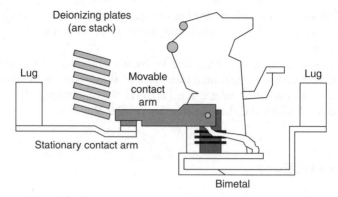

Figure 9.21 Thermal-magnetic circuit breaker.

9.5.5 Circuit Breakers

The circuit breaker is an electrical switch designed for the purpose of EOC events, short circuits, or fault detection [22]. Circuit breakers are typically "tripped" by the high-current event and can be manually reset. A pilot device senses a fault current and operates a trip-opening mechanism. The trip solenoid then releases a latch. Some high-voltage circuit breakers are self-contained with current transformers, protection relays, and an internal control power source. With detection of an electrical fault, the circuit breaker contacts open to interrupt the circuit; some mechanically stored energy contained within the breaker is used to separate the contacts. The circuit breaker can be reset manually after the event is over. Circuit breakers are nondestructive, as opposed to fuses which may only have a single use. Today, both fuses and circuit breakers can be integrated into a common system.

A class of circuit breakers is the thermal-magnetic circuit breaker. Thermal-magnetic circuit breakers are used to avoid "short-circuit" currents. Thermal-magnetic circuit breakers are sensitive to temperature.

Figure 9.21 shows an example of a thermal-magnetic circuit breaker [1]. Thermal-magnetic circuit breakers contain a bimetal switch and an electromagnet. The bimetal switch provides overcurrent protection. During current overload, the bimetal switch heats up, leading to bending of the element.

The electromagnet responds to short-circuit currents. The electromagnetic is a wire coil and an iron core. As a high current goes through the coil, induced magnetic field attracts an armature of the thermal-magnetic circuit breaker. When the armature extends toward the electromagnet due to the magnetic force, the armature rotates the trip bar. With the tripping of the trip bar, the current path is "open," and circuit breaker prevents the EOC.

9.6 EOS AND EMI PREVENTION: PRINTED CIRCUIT BOARD DESIGN

PCB design can influence the EOS robustness and its sensitivity to EMI. Today, PCB design is done with electronic design automation (EDA) tools. The EDA PCB design process undergoes the following steps [1]:

- **Circuit schematic:** Schematic capture is achieved using an EDA.

- **Card dimension:** Card dimension and template size are determined for the application. This is based on the component sizes and heat sinks.

- **Number of design levels:** The number of the design layers is determined based on designer's designation or requirements. Decisions include the following:

  o Power plane

  o Ground plane

  o Signal plane

- **Trace line impedance:** Decisions are made of the dielectric layer thickness copper routing thickness and trace width.

- **Component placement:** The placement of the components on the PCB is decided based on power distribution and physical dimensions.

- **Signal trace routing:** Signal lines are created using a "place and route" function.

- **File generation:** A Gerber file is generated for the final design.

9.6.1 Printed Circuit Board Power Plane and Ground Design

In the design of the PCB, a number of the design steps can influence the EOS, the EMI, and the electromagnetic compatibility (EMC) characteristics [1]:

- Card dimension

- Number of design levels

- Power plane design

- Ground plane design

- Signal plane design

- Trace line impedance

- Copper routing thickness

- Trace width

- Component placement

- Signal trace routing

Within this subject, guideline can be created to minimize the systems' sensitivities.

9.6.2 Printed Circuit Board Design Guidelines: Component Selection and Placement

As noted in the prior section, there are many decisions in the design of PCB that influence the EOS, EMI, and EMC sensitivities. The component selection and placement influences these issues. Some guidelines for printed circuit design for placement and component selection are given below [1]:

- **Component placement and ESD return current:** Components should avoid being placed over ESD return current ground traces.
- **Connector edge placement:** Connectors should be located on the same edge of a PCB.
- **Connector corner placement:** Connectors should be located on one corner of a PCB if possible.
- **Common connector:** All off-board signals from a single device should be routed through a common connector.
- **Connector to on-board I/O component spacing:** Components connected to I/O nets, connectors, and off-board components should be located within 2 cm away from connectors.
- **Connector and I/O to on-board non-I/O component spacing:** Components not connected to I/O nets or connectors should be located 2 cm away from I/O nets and connectors.
- **Digital component off-chip transition timing:** Digital components should be selected to have a maximum off-chip rise and fall times.
- **Clock and clock oscillator placement:** Clock drivers should be located within proximity of clock oscillators.

9.6.3 Printed Circuit Board Design Guidelines: Trace Routing and Planes

Trace routing and power/ground plane decisions are key to avoid EMI, EMC, and EOS concerns. Some guidelines for printed circuit design for signal traces are given below [1]:

- **Power trace EOS width:** All power supply line traces should be of suitable width for EOS robustness objectives.
- **Ground trace EOS width:** All ground line traces should be of suitable width for EOS robustness objectives.
- **Signal trace EOS width:** All signal line traces should be of suitable width for EOS robustness objectives.

- **Critical signal trace placement:** Critical signal traces should be placed between power and ground planes. This avoids susceptibility of noise, EMI, and other sources.

- **Critical signal trace placement and ESD return current:** Critical signal traces should avoid being placed over ESD return current ground traces.

- **Non-I/O trace placement:** Traces which are not I/O should not be located between an I/O connector and a device that receives or sends signals using that connector.

- **Non-I/O trace placement and I/O components:** Signals with high-speed contents should not be placed under components that contain I/O components.

- **Signal trace and power plane separation:** No trace should be used for connection to the power plane.

- **Signal trace and ground plane separation:** No trace should be used for connection to the ground plane.

- **I/O to connector trace length:** Trace lengths from I/O to connectors should be minimized.

- **High-speed digital trace length:** Trace lengths for high-speed digital signals should be minimized.

- **Clock trace length:** Trace lengths for clocks should be minimized.

- **High-speed trace to board edge:** High-speed traces should be routed at least $2X$ from the board edge, where X is the distance between the trace and its return current.

- **Differential signal trace pairs:** Differential signal trace pair should be routed together and maintain same distances from adjacent shapes, objects, and solid planes.

Some guidelines for printed circuit design for power planes are given below:

- **Power plane and trace common layer:** Power planes and traces should be routed on the same layer.

- **Power plane–power return plane (ground) common layer:** All power planes that are referenced to the same power return plane should be routed on the same layer.

In the earlier section, a list of design rules, recommendations, and guidelines is shown. From the design aspect, one can develop sophisticated EDA tools to further evaluate the integrity of the design. These tools can address the following (Figure 9.22):

- Identify EMI sources and victims

- Identify critical paths

- Identify potential antennae issues

- Identify coupling mechanisms

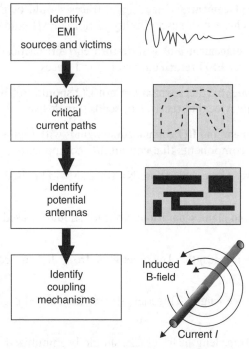

Figure 9.22 PCB design tool steps.

9.6.4 Printed Circuit Board Card Insertion Contacts

A commonly used technique to avoid system failures is to design card contacts so that the ground (V_{SS}) and power (V_{DD}) extend beyond the signal pins. The contacts of the power and ground are designed such that as the card is inserted into a system socket, contact occurs first with power and ground. The design is such that if the board (or card) was charged, the charge would flow to the system power or system ground first. Historically, there was a handheld device which did not design this correctly. When the handheld device was inserted into its socket to interface with a computer, a current pulse went into the computer signal lines, leading to computer failures.

9.6.5 System-Level Printed Circuit Board: Ground Design

Ground design is very critical in system board design [1]. In many board designs, the board designer separates the digital ground from the analog ground for system-level noise isolation. Signal lines on the board that cross the two ground planes are vulnerable to EMI on the signal lines.

A solution to address EMI and EMC problems is to avoid separation of the ground planes but have them connected at some location within the board [1]. Figure 9.23 shows an example of a digital and analog application, where the digital circuitry is in one section of the ground plane and the analog circuitry is in the other section of the same

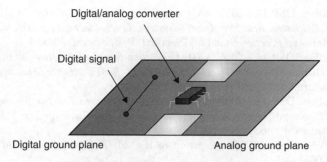

Digital/analog converter

Digital signal

Digital ground plane   Analog ground plane

Figure 9.23   Single ground board design.

ground plane. The ground plane has a small "bridge" where the digital-to-analog converter is placed. In this fashion, all the digital signal lines and pins remain on the one side of the plane; this avoids digital noise from impacting the analog circuitry but at the same time does not have a fully separated ground plane.

9.7 CLOSING COMMENTS AND SUMMARY

This chapter focused on the system-level issues associated with EOS in chips, PCB, and systems. EOS protection device classifications, symbols, and types for both overvoltage and overcurrent conditions are highlighted. System-level and system-like testing methods, such as IEC 61000-4-2, IEC 61000-4-5, and HMM waveforms and methods, were reviewed. Examples of printed circuit board (PCB) design practices for digital-to-analog systems were also shown.

In Chapter 10, latchup in analog design is discussed. Chapter 10 addresses solutions to avoid digital noise from impacting analog circuitry. Solutions such as spatial placement of digital and analog cores in a mixed-signal chip as well as guard rings between the domains are discussed. Moats, guard rings, and through-silicon via (TSV) advantages and disadvantages are discussed as possible solutions to minimize both noise and latchup are highlighted. Special features, such as grounded wells, and decoupling capacitor issues and how they can lead to latchup in analog applications are also reviewed.

REFERENCES

1. S. Voldman. *Electrical Overstress (EOS): Devices, Circuits, and Systems*. Chichester, UK: John Wiley & Sons, Ltd, 2013.
2. S. Voldman. *ESD Basics: From Semiconductor Manufacturing to Product Use*. Chichester, UK: John Wiley & Sons, Ltd, 2012.
3. S. Voldman. *ESD: Design and Synthesis*. Chichester, UK: John Wiley & Sons, Ltd, 2011.
4. S. Voldman. *Latchup*. Chichester, UK: John Wiley & Sons, Ltd, 2007.
5. V. Vashchenko and A. Shibkov. *ESD Design in Analog Circuits*. New York: Springer, 2010.
6. T. Meuse, R. Barrett, D. Bennett, M. Hopkins, J. Leiserson, J. Schichl, L. Ting, R. Cline, C. Duvvury, H. Kunz, and R. Steinhoff. Formation and suppression of a newly discovered secondary EOS event in HBM test systems. *Proceedings of the Electrical Overstress/Electrostatic Discharge (EOS/ESD) Symposium*, 2004; 141–145.

7. ESD Association. DSP 14.1-2003. *ESD Association Standard Practice for the Protection of Electrostatic Discharge Sensitive Items—System Level Electrostatic Discharge Simulator Verification Standard Practice.* Standard Practice (SP) document, 2003.

8. ESD Association. DSP 14.3-2006. *ESD Association Standard Practice for the Protection of Electrostatic Discharge Sensitive Items—System Level Cable Discharge Measurements Standard Practice.* Standard Practice (SP) document, 2006.

9. ESD Association. DSP 14.4-2007. *ESD Association Standard Practice for the Protection of Electrostatic Discharge Sensitive Items—System Level Cable Discharge Test Standard Practice.* Standard Practice (SP) document, 2007.

10. H. Geski. DVI compliant ESD protection to IEC 61000-4-2 level 4 standard. *Conformity,* September 2004; 12–17.

11. IEC 61000-4-2. Electromagnetic compatibility (EMC): Testing and measurement techniques— Electrostatic discharge immunity test. *IEC International Standard,* 2001.

12. E. Grund, K. Muhonen, and N. Peachey. Delivering IEC 61000-4-2 current pulses through transmission lines at 100 and 330 ohm system impedances. *Proceedings of the Electrical Overstress/Electrostatic Discharge (EOS/ESD) Symposium,* 2008; 132–141.

13. IEC 61000-4-2. Electromagnetic compatibility (EMC)—Part 4-2: Testing and measurement techniques—Electrostatic discharge immunity test. *IEC International Standard,* 2008.

14. IEC 61000-4-5. Electromagnetic compatibility (EMC)—Part 4-5: Testing and measurement techniques—Surge immunity test, *IEC International Standard,* 2000.

15. R. Chundru, D. Pommerenke, K. Wang, T. Van Doren, F.P. Centola, and J.S. Huang. Characterization of human metal ESD reference discharge event and correlation of generator parameters to failure levels—Part I: Reference event. *IEEE Transactions on Electromagnetic Compatibility,* **46** (4), November 2004; 498–504.

16. K. Wang, D. Pommerenke, R. Chundru, T. Van Doren, F.P. Centola, and J.S. Huang. Characterization of human metal ESD reference discharge event and correlation of generator parameters to failure levels—Part II: Correlation of generator parameters to failure levels. *IEEE Transactions on Electromagnetic Compatibility,* **46** (4), November 2004; 505–511.

17. ESD Association. ESD-SP5.6-2008. *ESD Association Standard Practice for the Protection of Electrostatic Discharge Sensitive Items—Electrostatic Discharge Sensitivity Testing—Human Metal Model (HMM) Testing Component Level.* Standard Practice (SP) document, 2008.

18. ANSI/ESD SP5.6-2009. *Electrostatic Discharge Sensitivity Testing—Human Metal Model (HMM)—Component Level,* 2009.

19. R. Ashton. *Types of Electrical Overstress Protection.* AND9009D, On Semiconductor Application Note, http://www.onsemi.com, May 2011; 1–13.

20. W. Schossig. Introduction to history of selective protection. *PAC Magazine,* 2007; 70–74.

21. R. Ashton and L. Lescouzeres. Characterization of off-chip ESD protection devices. *Proceedings of the Electrical Overstress/Electrostatic Discharge (EOS/ESD) Symposium,* 2008; 21–29.

22. C.G. Page. Improvement in induction-coil apparatus and in circuit breakers. U.S. Patent No. 76,654, April 14, 1868.

23. T.A. Edison. Fuse block. U.S. Patent No. 438,305, October 14, 1890.

24. A. Wright and P.G. Newbury. *Electric Fuses.* 3rd Edition. London, UK: Institute of Electrical Engineers, 2004.

10 Latchup Issues for Analog Design

10.1 LATCHUP IN ANALOG APPLICATIONS

CMOS latchup is a concern in advanced semiconductor CMOS, BiCMOS, and bipolar–CMOS–LDMOS (BCD) technologies within a given circuit, between adjacent circuits, and between domains [1–59]. Due to technology scaling, the physical distances between p-channel MOSFETs and n-channel MOSFETs continue to be reduced in the periphery and core of circuits. With density scaling, the number of I/O circuits increases according to Rent's rule. As a result, the aspect ratio of peripheral I/O circuitry continues to move toward "long/narrow I/O standard cells" with decreased spacing between adjacent I/O standard cells. Hence, the interaction between adjacent I/O (e.g., I/O to I/O) will continue to be a design issue associated with CMOS latchup. In addition, with mixed signal (MS) and system on chip (SOC), the placement of circuits of different domains can also lead to CMOS latchup concerns.

Figure 10.1 shows possible latchup issues in a semiconductor chip. The focus in this chapter will be on the issue of I/O-to-I/O latchup [6, 7], guard rings [8–18], through-silicon via (TSV) [32–39], deep trench [40–46], and active guard rings [47–59]. Test structures that address I/O-to-I/O interactions will be discussed. Electrical measurements of parasitic bipolar current gain and analysis will be shown.

10.2 I/O-TO-I/O LATCHUP

I/O-to-I/O latchup can occur with the formation of a lateral parasitic pnpn network between two adjacent I/O cells [6]. With two adjacent I/O standard cells, latchup can occur in multiple interactions due to the two PFETs and two NFETs. The adjacent standard cells can be the following:

- Digital-to-analog I/O
- Analog I/O to analog I/O

ESD: Analog Circuits and Design, First Edition. Steven H. Voldman.
© 2015 John Wiley & Sons, Ltd. Published 2015 by John Wiley & Sons, Ltd.

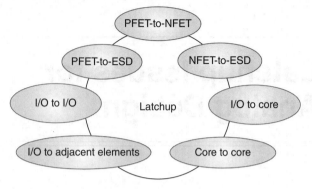

Figure 10.1 Latchup.

- High-voltage I/O to analog I/O
- Analog I/O to ESD
- Analog I/O to ESD power clamp
- Analog I/O to grounded n-well plate
- Analog I/O to decoupling capacitor

I/O-to-I/O latchup can occur between two n-well regions of adjacent analog standard cell and an adjacent structure. To quantify I/O-to-I/O latchup concerns, a structure can be formed to address all interactions. Two four-stripe structures can be formed in a test structure as shown in Figure 10.2.

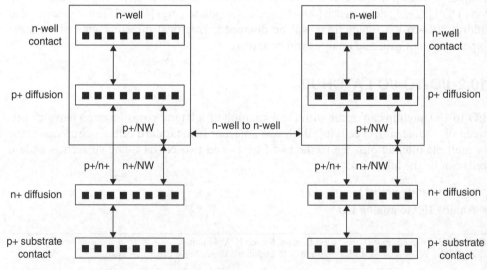

Figure 10.2 Standard cell I/O-to-I/O latchup test structure.

10.3 I/O-TO-I/O LATCHUP: N-WELL TO N-WELL

I/O-to-I/O latchup can occur between two n-well regions of adjacent analog standard cells. A lateral pnpn can be formed between the first I/O network PFET, its own n-well, the p-substrate, and an adjacent n-well (Figure 10.3) [6].

The critical parameter of interest is the spacing between the two adjacent cells forming a lateral npn between the first and second n-well regions. As the technologies are scaled, the bond pad will be scaled, decreasing the space between the two adjacent n-well regions. As the spacing decreases, the npn bipolar current gain will increase with the n-well-to-n-well space. Note that this interaction is symmetrical and bidirectional (e.g., there are two pnpn parasitic elements formed). Tables 10.1 and 10.2 show the lateral bipolar npn gain as a function of n-well-to-n-well spacing at 25 and 125°C, respectively (Figures 10.4 and 10.5).

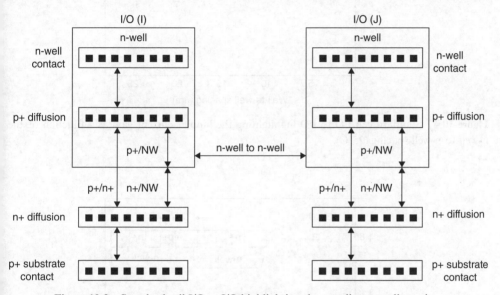

Figure 10.3 Standard cell I/O to I/O highlighting the n-well-to-n-well spacing.

Table 10.1 Lateral NW–NW bipolar current gain (25°C)

Structure	Well to well (μm)	NW–NW npn beta
I/O to I/O	5	0.6
I/O to I/O	10	0.5
I/O to I/O	15	0.3
I/O to I/O	20	0.2

Table 10.2 Lateral NW–NW bipolar current gain (125°C)

Structure	Well to well (μm)	NW–NW npn beta
I/O to I/O	5	1.8
I/O to I/O	10	1.6
I/O to I/O	15	1.45
I/O to I/O	20	1.3

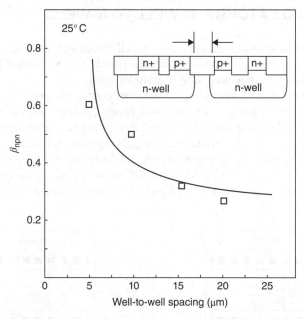

Figure 10.4 Standard cell I/O to I/O highlighting the bipolar current gain as a function of the n-well-to-n-well spacing (25°C).

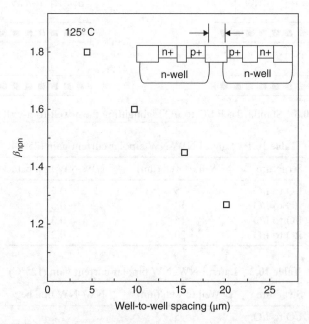

Figure 10.5 Standard cell I/O to I/O highlighting the bipolar current gain as a function of the n-well-to-n-well spacing (125°C).

10.4 I/O-TO-I/O LATCHUP: N-WELL TO NFET

A lateral pnpn can be also formed between the first I/O network PFET, its own n-well, the p-substrate, and the adjacent n-channel MOSFET device of the second I/O cell (Figure 10.6) [6]. In this case, the NFET pull-down source serves as the emitter of the pnpn network. Test structures varied all the design variables to evaluate the npn bipolar current gain between the n-well of the first standard cell and the adjacent "NFET" structure.

Table 10.3 contains experimental results of NW to n+ where the well-to-well spacing was varied from 15 to 5 μm (Figures 10.7 and 10.8). Table 10.3 is data for a p+/n+ spacing of 2.4 μm.

An interesting experimental results showed that as the NFET (I/O = 2) became closer to the n-well (I/O = 2), the bipolar current gain between the n-well (I/O = 1) and NFET (I/O = 2) *decreased*!.

Table 10.4 shows the bipolar gain of NW(1) –to – n+ (2), as the p+/n+ spacing is reduced in the design. The lateral bipolar current gain decreases from 0.5 to 0.4 (at 25°C); note that it decreased instead of increasing. This is counterintuitive. The reason for the reduction is the NW(2) starts to collect the carrier from NW(I/O = 1) to n + (I/O = 2), serving as a "pseudo guard ring" for this interaction [6].

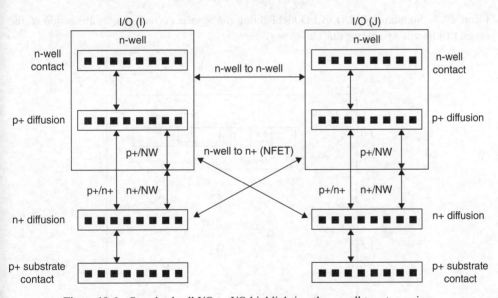

Figure 10.6   Standard cell I/O to I/O highlighting the n-well to n+ spacing.

Table 10.3   Lateral NW–n+ bipolar current gain as a function of well to well (25°C)

Structure	Well to well (μm)	NW–n+ npn beta
I/O to I/O	5	0.5
I/O to I/O	10	0.39
I/O to I/O	15	0.3

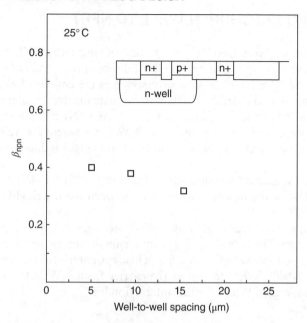

Figure 10.7   Standard cell I/O to I/O highlighting the bipolar current gain as a function of the n-well to adjacent n+ spacing (25°C).

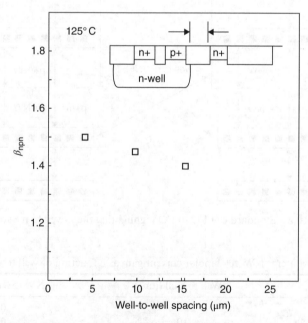

Figure 10.8   Standard cell I/O to I/O highlighting the bipolar current gain as a function of the n-well to adjacent n+ spacing (125°C).

Table 10.4   Lateral NW–n+ bipolar current gain (25°C) versus p+/n+ space

p+/n+ (μm)	Well to well (μm)	NW–n+ npn beta
2.4	5	0.5
1.2	5	0.44
0.8	5	0.4

10.5 I/O-TO-I/O LATCHUP: NFET TO NFET

A third interaction of interest is the lateral npn bipolar transistor formed between the two adjacent NFET devices [6]. In this case, only a lateral npn is formed (Figure 10.9). Table 10.5 shows the experimental results as a function of the n-well-to-n-well spacing (leading to a smaller base width between the two NFETs).

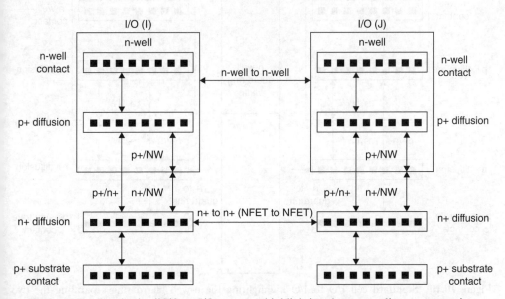

Figure 10.9   Standard cell I/O-to-I/O structure highlighting the n+ to adjacent n+ spacing.

Table 10.5   Lateral NFET–NFET current gain (125°C)

Structure	Well to well (μm)	n+ to n+ npn beta
I/O to I/O	5	0.28
I/O to I/O	10	0.22
I/O to I/O	15	0.20
I/O to I/O	20	1.3

10.6 I/O-TO-I/O LATCHUP: N-WELL GUARD RING BETWEEN ADJACENT CELLS

In some foundries, an n-well guard ring is placed between the two adjacent cells. With the placement of the n-well guard ring, carriers will be collected by the adjacent n-well guard ring instead of the adjacent I/O cell's n-well or NFET region [6]. Figure 10.10 shows the standard cell with an n-well guard ring diffusion, and Table 10.6 shows the experimental results. In this implementation, the issue is the npn bipolar current gain between the PFET tub and the n-well guard ring, as well as the n-well guard ring series resistance.

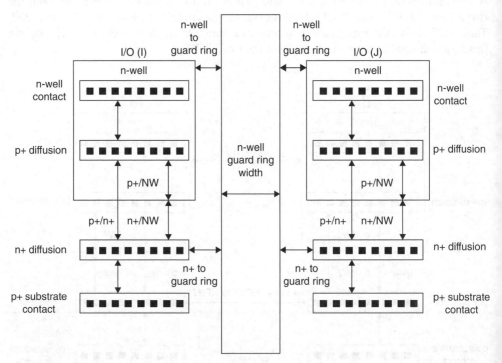

Figure 10.10 Standard cell I/O to I/O highlighting the n-well guard ring separating the two adjacent I/O circuits.

Table 10.6 Lateral NW-to-NW guard ring current gain (25°C)

Structure	Well-to-guard ring (μm)	n+ to n guard ring npn beta
I/O to I/O	5	6.0
I/O to I/O	10	5.0
I/O to I/O	15	4.0
I/O to I/O	20	3.4

10.7 LATCHUP OF ANALOG I/O TO ADJACENT STRUCTURES

Many CMOS latchup problems today occur due to the placement of circuits adjacent to each other without verification of the interaction. This will become more important in analog design as technology spacings are reduced in CMOS, BiCMOS, BCD, and high-voltage and ultrahigh-voltage technologies.

10.7.1 Latchup in Core-Dominated Semiconductor Chips

Many CMOS latchup problems in analog design occur due to low pin count in a chip periphery where extra space exists. In these low pin count analog applications, the designs are "core dominated" (as opposed to peripheral I/O dominated in digital applications). Extra space exists in the chip periphery, where non-I/O circuitry is placed. CMOS latchup can occur between the analog I/O standard or custom cells and these other structures adjacent to the I/O networks.

10.7.2 Latchup and Grounded N-Wells

A common problem occurs when CMOS circuits are placed near grounded n-well structures. Many CMOS latchup problems today occur due to the placement of circuits adjacent to grounded n-well regions without verification of the interaction [5, 6]. Grounded n-well structures and grounded n+ diffusions near off-chip driver PFET structures have historically caused latchup failures [5].

10.7.3 Latchup and Decoupling Capacitors

In analog design, the density of the I/O is not as limited due to the reduced pin count for analog applications [6, 7]. One of the critical problems is that digital standard cells are used for analog applications, where it is anticipated that the periphery is 100% I/O cells. Figure 10.11 shows an example where the decoupling capacitors were placed between the standard cell I/O circuitries. The decoupling capacitor contains a grounded n-well bottom plate to form the capacitor structure. With the grounded n-well plate adjacent to an I/O PFET, a lateral pnpn is formed between the PFET, the n-well associated with the PFET, the substrate, and the decoupling capacitor.

10.7.4 Adjacency Design Rule Checking and Verification

CMOS latchup occurs in analog and MS application due to the lack of design rule checking (DRC) and LVS checking and verification to address all possible cases of interactions. Adjacency rules can be established to check and verify any lateral issues to structures not previously used near I/O cells.

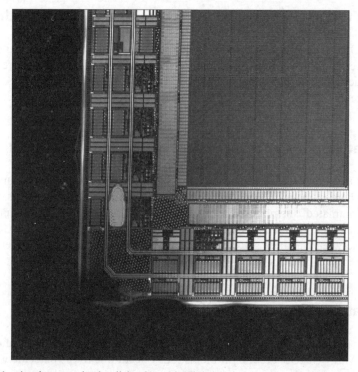

Figure 10.11   Analog standard cell latchup highlighting interaction between standard cell I/O PFET and adjacent decoupling capacitor element.

10.8 ANALOG I/O TO CORE

CMOS latchup occurs in analog and MS application between the I/O circuitry and the cores [3]. Guard rings can be placed to separate the analog I/O from digital cores and the digital I/O from the analog cores. Figure 10.12 shows test structures to evaluate external latchup between I/O and core circuitries. The placement of I/O PFET and I/O NFET "injectors" from core circuitry is a design issue. In addition, guard rings can be placed between the I/O and the core networks to collect minority injection carriers. These guard rings can also help reduce the noise interaction between the I/O and analog cores.

10.9 CORE-TO-CORE ANALOG–DIGITAL FLOOR PLANNING

In a semiconductor chip, the analog and digital core floor planning is key to prevent the digital noise from impacting analog circuitry. Spatial separation of the circuits and signal lines is critical as well as isolation using moats and guard rings.

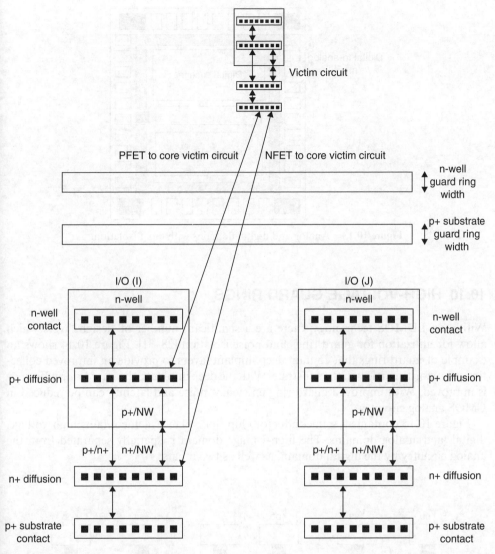

Figure 10.12   I/O to core test structure for evaluation of external latchup between I/O and core networks.

10.9.1 Analog–Digital Moats and Guard Rings

Within a MS chip, analog and digital domains are physically separated to avoid digital noise and injection from affecting analog circuitry. Figure 10.13 is an example of MS floor plan. The digital and analog domains are separated by establishing a "moat." The physical spacing between the digital and analog domains can be moat widths on the order of 10–50 µm.

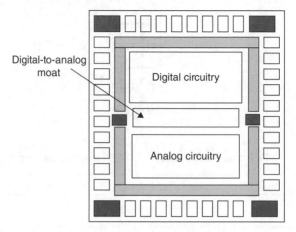

Figure 10.13   Analog and digital domains with moat isolation.

10.10 HIGH-VOLTAGE GUARD RINGS

Within an LDMOS technology, there are a significant number of design layers which allow for utilization for guard rings and noise isolation [28–31]. Figure 10.14 shows an example of guard rings that contain deep implant layers to provide an improved collection of minority carriers in the substrate. With the deep implants, the guard ring efficiency is improved. With improved guard ring efficiency, noise and latchup can be reduced in CMOS analog circuitry.

Figure 10.15 contains a semiconductor chip floor plan which contains high-voltage, digital, and analog domains. The high-voltage domain is spatially separated from the analog circuitry by the digital domain, as well as two moats.

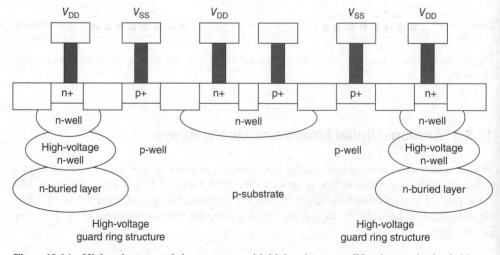

Figure 10.14   High-voltage guard ring structure with high voltage n-well implant and n-buried layer.

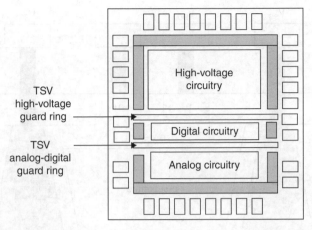

Figure 10.15   Floorplan for high voltage, digital and analog function with through silicon via (TSV).

10.11 THROUGH-SILICON VIA (TSV)

TSV structures have been introduced in advanced technologies for multichip integration. TSV structures have been integrated into silicon chips and silicon interposers. In MS chips, where digital and analog functions are on the same semiconductor chip, TSV can be used between the two domains to provide noise isolation and minimize minority carrier injection [32–39]. Figure 10.16 shows an example of a TSV guard ring structure. Figure 10.17 shows a cross section of the wafer highlighting the TSV depth extends from the top to the bottom of the wafer. TSV structures can be placed in the moat area between the digital and analog domains, allowing collection of the minority carriers and eliminating external latchup.

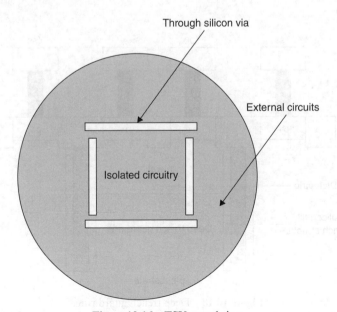

Figure 10.16   TSV guard ring.

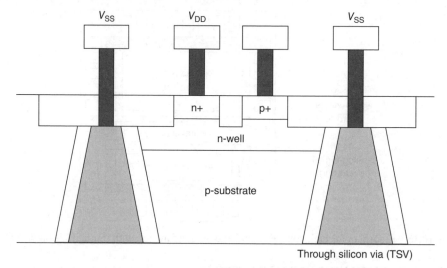

Figure 10.17   Wafer cross section of a TSV.

10.12 TRENCH GUARD RINGS

Deep trench structures can be utilized to improve the latchup robustness and noise reduction in CMOS, BiCMOS, and LDMOS technologies [40–46]. Deep trench structures are used for high-performance bipolar transistors. Deep trench structures reduce collector to substrate capacitance in bipolar transistors; these can be utilized for noise reduction and latchup. Deep trench structures can be used for guard rings (Figures 10.18 and 10.19). Deep trench structures can be on the order of 5–10 μm deep from the silicon surface. Deep trench structures prevent lateral minority carrier transport.

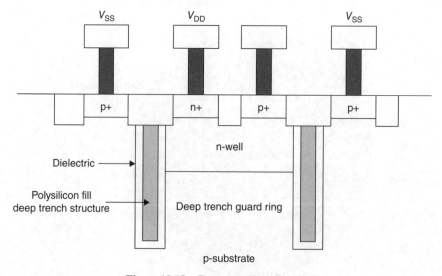

Figure 10.18   Deep trench guard ring.

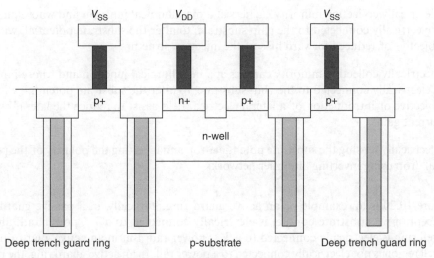

Figure 10.19   Independent guard ring structure.

10.13 ACTIVE GUARD RINGS

Today, SOC solutions have been used for solving the MS requirements. SOC applications have a wide range of power supply conditions, number of independent power domains, and circuit performance objectives. Different power domains are established between digital and analog circuits on an integrated chip. The integration of different circuits and system functions into a common chip has also resulted in solutions for ensuring that noise from one portion or circuit of the chip does not affect a different circuit within the chip.

With the chip integration issues, the need for better guard rings and alternative guard ring solutions has had increased interest [47–59]. Since 2000, there has been an increased focus on guard ring solutions that achieve the following objectives:

- Solutions that achieve noise isolation, latchup robustness, and ESD results
- Solutions which do not inject current back into the power grid

With the growth of interaction between digital and analog domains, new guard ring concepts have increased in importance. In addition, with the growth of smart power technology, solutions are needed for avoidance of interaction of the high-voltage CMOS (HVCMOS) chip sectors and the low-voltage sectors of a CMOS chip.

Different "active" guard ring circuit concepts have been introduced for latchup improvement. In "active" guard rings, the objective is to not only collect minority carriers but to actively compensate the effect. The latchup circuit design discipline includes the following concepts:

- Electrically collecting minority carriers at a metallurgical junction and whose junction is electrically connected to the chip substrate to alter the substrate potential

- Electrically collecting minority carriers at a metallurgical junction and whose junction is electrically connected to the chip substrate, to alter the substrate potential, with the objective of reduced forward bias of the injection structure

- Electrically collecting minority carriers at a metallurgical junction and whose junction is electrically connected to the chip substrate, to alter the substrate potential, with the objective of introduction of a lateral electrical field assist to reduce the lateral bipolar current gain

- Electrically sensing the substrate potential drop and inverting the polarity of the potential drop using inverting amplifier networks

Figure 10.20 is an example of an active guard ring. Typically, in a passive guard ring concept, a p+ substrate contact is electrically connected to a V_{SS} power rail, and an n-well ring is electrically connected to a V_{DD} power rail. But in an active guard ring, an n-well region is not electrically connected to a power rail. In an active guard ring, the n-well structure collects the minority carrier electrons in its metallurgical junction formed with the p-substrate region. The n-well ring is electrically connected to a "soft grounded" p+ substrate contact. When the minority carrier electrons traverse the metallurgical junction, it reduces the electrical potential of the n-well region (e.g., denoted in the figure as ΔV). By electrically connecting the n-well to the p+ substrate contact, the electrical potential of the substrate is also lowered by the same potential magnitude. In this case, the electrical potential of the region is lowered. The lowering of the substrate potential can be utilized as two means. First, given the p+ diffusion is near a forward-bias structure (e.g., an injecting structure), the reduction of the potential can lower the forward-bias state, turning off the injection process. Second, given another p+ substrate contact, a lateral electric field can be established which inhibits the flow of minority carriers. Given a parasitic npn bipolar transistor is formed between the injection source and a collecting victim circuit, if the lateral electrical field opposes the current transport, the lateral npn bipolar current gain is reduced.

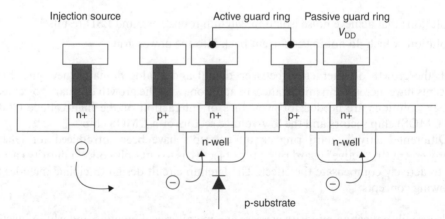

Figure 10.20   N+ injection source with active and passive guard rings.

Figure 10.21 demonstrates the "electric field assist" wherein in this case, the electric field reduces the lateral bipolar current gain. In this methodology, the placement of the p+ region can be on the injection side or collection side of the n-well region. By adding an additional p++ substrate diffusion inside of the n-well ring/p+ substrate contact, a well defined electric field is established. A p+ region is flanking both sides of the n-well ring, with an outer p+ substrate contact electrically connected to the n-well ring.

Figure 10.22 introduces a secondary p+ passive guard ring. Various implementations of guard rings are utilized where a plurality of p+ substrate contacts and n-wells are integrated,

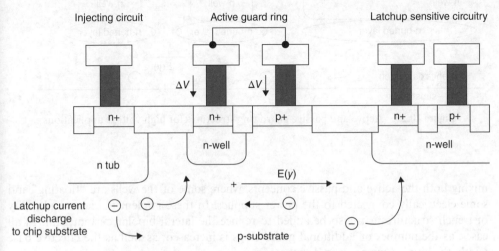

Figure 10.21   N-well Injection source with active guard ring.

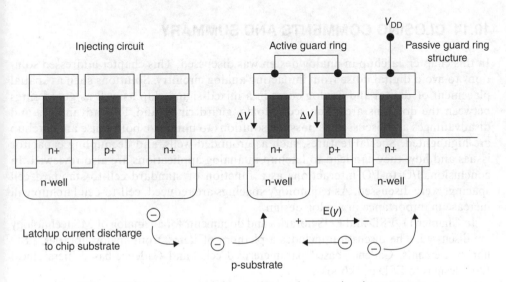

Figure 10.22   Active guard rings and secondary passive ring structure.

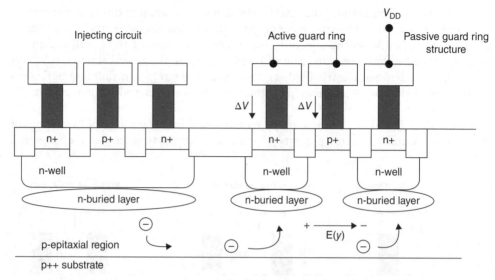

Figure 10.23   Active and passive guard ring structures for high voltage applications.

mixing both the active and passive concepts, where some of the wells are "floating" and some electrically connected to the power supplies. In these implementations, a plurality of trench structures can also be added to reduce the lateral bipolar current gain. In all cases, as the number of additional guard rings is increased, as well as the effective base width, the bipolar current gain decreases.

10.14 CLOSING COMMENTS AND SUMMARY

In this chapter, latchup in analog design was discussed. This chapter addressed solutions to avoid digital noise from impacting analog circuitry. Solutions such as spatial placement of digital and analog cores in a mixed-signal chip as well as guard rings between the domains are discussed. Moats, guard rings, and TSV advantages and disadvantages are discussed as possible solutions to minimize both noise and latchup are highlighted. Special features, such as grounded wells, and decoupling capacitor issues and how they can lead to latchup in analog applications are also reviewed. In conclusion, I/O-to-I/O interactions as a function of standard cell-to-standard cell spacings were discussed. As technology spacings are reduced, cell-to-cell latchup will increase in importance in analog design.

In Chapter 11, ESD and EOS libraries and documents for an analog or MS technology are discussed. The discussion includes a plethora of items, from analog libraries, ESD library elements, Cadence-based parameterized cells, and Cadence-based hierarchical ESD designs to ESD cookbooks.

REFERENCES

1. R. Troutman. *Latchup in CMOS Technology: The Problem and Its Cure.* New York: Springer, 1985.
2. S. Voldman. *Latchup.* Chichester, UK: John Wiley & Sons, Ltd, 2007.
3. M.D. Ker and S.F. Hsu. *Transient-Induced Latchup in CMOS Integrated Circuits.* Hoboken, NJ: John Wiley & Sons, Inc., 2009.
4. S. Voldman. *Electrical Overstress (EOS): Devices, Circuits, and Systems.* Chichester, UK: John Wiley & Sons, Ltd, 2013.
5. Y. Huh, K. Min, P. Bendix, V. Axelrad, R. Narayan, J.W. Chen, L.D. Johnson, and S. Voldman. Chip level layout and bias considerations for preventing neighboring I/O cell interaction-induced latchup and inter-power supply latchup in advanced CMOS technologies. *Proceedings of the Electrical Overstress/Electrostatic Discharge (EOS/ESD) Symposium*, 2005; 100–107.
6. S. Voldman. Latchup I/O to I/O adjacency issues in peripheral I/O design for digital and analog applications. *Proceedings of the International Conference Semiconductor and Circuit Technology (ICSICT)*, 2012.
7. S. Voldman. ESD and latchup considerations for analog and power applications. *Proceedings of the International Conference Semiconductor and Circuit Technology (ICSICT)*, 2012.
8. R.R. Troutman. Epitaxial layer enhancement of n-well guard rings for CMOS circuits. *IEEE Electron Device Letters*, **4** (12), 1983; 438–440.
9. J. Quinke. Novel test structures for the investigation of the efficiency of guard rings used for I/O latchup prevention. *Proceedings on the International Conference on Microelectronic Test Structures (ICMTS)*, 1990; 35–40.
10. D. Tremouilles, M. Bafluer, G. Bertrand, and G. Nolhier. Latch-up ring design guidelines to improve electrostatic discharge (ESD) protection scheme efficiency. *IEEE Journal of Solid-State Circuits*, **39** (10), 2005; 1778–1782.
11. S. Voldman, C.N. Perez, and A. Watson. Guard rings: Theory, experimental quantification, and design. *Proceedings of the Electrical Overstress/Electrostatic Discharge (EOS/ESD) Symposium*, October 2005; 131–140.
12. S. Voldman, C.N. Perez, and A. Watson. Guard rings: Structures, design methodology, integration, experimental results, and analysis for RF CMOS and RF mixed signal silicon germanium technology. *Journal of Electrostatics*, **64**, 2006; 730–743.
13. D. Tremouilles, M. Scholz, G. Groseneken, M.I. Natarajan, N. Azilah, M. Bafluer, M. Sawada, and T. Hasebe. A novel method for guard ring efficiency assessment and its applications for ESD protection design and optimization. *Proceedings of the International Reliability Physics Symposium (IRPS)*, 2007; 606–607.
14. T. Cavioni, M. Cecchetti, M. Muschitiello, G. Spiazzi, I. Vottre, and E. Zanoni. Latch-up characterization in standard and twin-tub test structures by electrical measurements, 2-D simulations and IR microscopy. *Proceedings on the International Conference on Microelectronic Test Structures (ICMTS)*, 1990; 41–46.
15. C. Mazure, W. Reczek, D. Takacs, and J. Winnerl. Improvement of latching hardness by geometry and technology tuning. *IEEE Transactions on Electron Devices*, **ED-35** (10), 1988; 1609–1615.
16. Y. Song, J.S. Cable, K.N. Vu, and A.A. Witteles. The dependence of latchup sensitivity on layout features in CMOS integrated circuits. *IEEE Transactions of Nuclear Science*, **NS-33** (6), 1986; 1493–1498.
17. R. Lohia and A. Ali. Parametric formulation of CMOS latchup as a function of chip layout parameters. *IEEE Journal of Solid State Circuits*, **23** (1), February 1988; 245–250.

18. R. Menozzi, L. Selmi, E. Sangiorgi, G. Crisenza, T. Cavioni, and B. Ricco. Layout dependence of CMOS latchup. *IEEE Transactions on Electron Devices*, **ED-35** (11), 1988; 1892–1901.

19. M.D. Ker and J.J. Peng. Layout design and verification for cell library to improve ESD/latchup reliability in deep-submicron CMOS technology. *IEEE Custom Integrated Circuit Conference (CICC)*, 1998; 537–540.

20. T. Aoki. A practical high-latchup immunity design methodology for internal circuits in the standard cell-based CMOS/BiCMOS LSI's. *IEEE Transactions on Electron Devices*, **40** (8), August 1993; 1432–1436.

21. B. Basaran, R.A. Rutenbar, and L.R. Carley. Latchup-aware placement and parasitic-bounded routing of custom analog cells. *IEEE International Conference on Computer Aided Design*, 1993; 415.

22. H. de La Rochette, G. Bruguier, J.M. Palau, and J. Gasiot. The effect of layout modification on latchup triggering in CMOS by experimental and simulation approaches. *IEEE Transaction on Nuclear Science*, **41** (6), December 1994; 2222.

23. S. Bhattacharya, S. Banerjee, J. Lee, A. Tasch, and A. Chatterjee. Design issues for achieving latchup-free, deep trench-isolated, bulk, non-epitaxial, submicron CMOS. *Proceedings of the International Electron Device Meeting (IEDM)*, 1990; 185–188.

24. M.D. Ker, W.Y. Lo, and C.Y. Wu. New experimental methodology to extract compact layout rules for latchup prevention in bulk CMOS IC's. *Proceedings of the IEEE Custom Integrated Circuits Conference (CICC)*, 1999; 143–146.

25. M.D. Ker and J.J. Peng. Layout design and verification for cell library to improve ESD/latchup reliability in deep-submicron CMOS technology. *Proceedings of the IEEE Custom Integrated Circuits Conference (CICC)*, 1998; 537–540.

26. S. Voldman. Methodology for placement based on circuit function and latchup sensitivity. U.S. Patent No. 8,108,822, January 31, 2012.

27. S. Voldman. Latch-up analysis and parameter modification. U.S. Patent No. 6,996,786, February 7, 2006.

28. S. Voldman. Structure, structure and method of latch-up immunity for high and low voltage integrated circuits. U.S. Patent No. 8,519,402, August 27, 2013.

29. S. Voldman. Semiconductor structure and method of designing semiconductor structure to avoid high voltage initiated latch-up in low voltage sectors. U.S. Patent No. 8,423,936, August 27, 2013.

30. S. Voldman. Guard ring structures for high voltage CMOS/low voltage CMOS technology using LDMOS (lateral double-diffused metal oxide semiconductor) device fabrication. U.S. Patent No. 8,110,853, February 7, 2012.

31. S. Voldman. Structure and method for latchup improvement using wafer via latchup guard ring. U.S. Patent No. 7,989,282, August 2, 2011.

32. S. Voldman. Structure and method for latchup improvement using wafer via latchup guard ring. U.S. Patent No. 8,390,074, March 5, 2013.

33. P. Chapman, D.S. Collins, and S. Voldman. Structure and method for latchup robustness with placement of through wafer via within CMOS circuitry. U.S. Patent No. 8,420,518, April 16, 2013.

34. S. Voldman. ESD network circuit with a through wafer via structure and a method of manufacture. U.S. Patent No. 8,232,625, July 31, 2012.

35. P. Chapman, D.S. Collins, and S. Voldman. Structure and method for latchup robustness with placement of through wafer via within CMOS circuitry. U.S. Patent No. 8,017,471, September 13, 2011.

36. P. Chapman, D.S. Collins, and S. Voldman. Structure for a latchup robust array I/O using through wafer via. U.S. Patent No. 7,855,420, December 21, 2010.

37. P. Chapman, D.S. Collins, and S. Voldman. Latchup robust array I/O using through wafer via. U.S. Patent No. 7,741,681, June 22, 2010.

38. P. Chapman, D.S. Collins, and S. Voldman. Structure for a latchup robust gate array using through wafer via. U.S. Patent No. 7,696,541, April 13, 2010.

39. P. Chapman, D.S. Collins, and S. Voldman. Latchup robust gate array using through wafer via. U.S. Patent No. 7,498,622, March 3, 2009.

40. A. Watson, S. Voldman, and T. Larsen. Deep trench guard ring structures and evaluation of the probability of minority carrier escape for ESD and latchup in advanced BiCMOS SiGe technology. *Proceedings of the Taiwan Electrostatic Discharge Conference (T-ESDC)*, 2003; 97–103.

41. S. Voldman and A. Watson. The influence of deep trench and substrate resistance on the latchup robustness in a BiCMOS silicon germanium technology. *Proceedings of the International Reliability Physics Symposium (IRPS)*, 2004; 135–142.

42. S. Voldman and A. Watson. The influence of polysilicon-filled deep trench and sub-collector implants on latchup robustness in RF CMOS and BiCMOS SiGe technology. *Proceedings of the Taiwan Electrostatic Discharge Conference (T-ESDC)*, 2004; 15–19.

43. S. Voldman. The influence of a novel contacted polysilicon-filled deep trench (DT) biased structure and its voltage bias state on CMOS latchup. *Proceeding of the International Reliability Physics Symposium (IRPS)*, 2006; 151–158.

44. C.N. Perez and S. Voldman. Method of forming a guard ring parameterized cell structure in a hierarchical parameterized cell design, checking and verification system. U.S. Patent No. 7,350,160, March 25, 2008.

45. C.N. Perez and S. Voldman. Method of displaying a guard ring within an integrated circuit. U.S. Patent No. 7,350,160, March 25, 2008.

46. V. Parthasarathy, R. Zhu, V. Khemka, T. Roggenbauer, A. Bose, P. Hui, P. Rodriguez, J. Nivison, D. Collins, Z. Wu, I. Puchades, and M. Butner. A 0.25-μm CMOS based 70 V smart power technology with deep trench for high-voltage isolation. *International Electron Device Meeting (IEDM) Technical Digest*, 2002; 459–462.

47. M. Bafluer, J. Buxo, M.P. Vidal, P. Givelin, V. Macary, and G. Sarrabayrouse. Application of a floating well concept to a latchup-free low-cost smart power high-side switch technology. *IEEE Transactions on Electron Devices*, **ED-40** (7), July 1993; 1340–1342.

48. R. Peppiette. A new protection technique for ground recirculation parasitics in monolithic power IC's. *Sanken Technical Report*, **26** (1), 1994; 91–97.

49. M. Bafluer, M.P. Vidal, J. Buxo, P. Givelin, V. Macary, and G. Sarrabayrouse. Cost-effective smart power CMOS/DMOS technology: design methodology for latchup immunity. *Analog Integrated Circuits and Signal Processing*, **8** (3), November 1995; 219–231.

50. W.W.T. Chan, J.K.O. Sin, and S.S. Wong. A novel crosstalk isolation structure for bulk CMOS power IC's. *IEEE Transactions on Electron Devices*, **ED-45** (7), July 1998; 1580–1586.

51. W. Winkler and F. Herzl. Active substrate noise suppression in mixed-signal circuits using on-chip driven guard rings. *Proceedings of the IEEE 2000 Custom Integrated Circuits Conference*, May 2000; 356–360.

52. O. Gonnard and G. Charitat. Substrate current protection in smart power IC's. *Proceedings of the International Symposium on Power Semiconductor Devices (ISPSD)*, 2000; 169–172.

53. R. Zhu, V. Parthasarathy, V. Khemka, and A. Bose. Implementation of high-side, high-voltage RESURF LDMOS in a sub-half micron smart power technology. *Proceedings of the International Symposium on Power Semiconductor Devices (ISPSD)*, 2001; 403–406.

54. O. Gonnard, G. Charitat, P. Lance, M. Susquet, M. Bafluer, and J.P. Laine. Multi-ring active analogic protection (MAAP) for minority carrier injection suppression in smart power IC's. *Proceedings of the International Symposium on Power Semiconductor Devices (ISPSD)*, 2001; 351–354.

55. M. Schenkel, P. Pfaffli, W. Wilkening, D. Aemmer, and W. Fichtner. Transient minority carrier collection from substrate in smart power design. *Proceedings of the European Solid State Device Research Conference (ESSDERC)*, 2001; 411–414.

56. V. Parthasarathy, V. Khemka, R. Zhu, I. Puchades, T. Roggenbauer, M. Butner, P. Hui, P. Rodriquez, and A. Bose. A multi-trench analog + logic protection (M-TRAP) for substrate crosstalk prevention in a 0.25-μm smart power platform with 100 V high-side capability. *Proceedings of the International Symposium on Power Semiconductor Devices (ISPSD)*, 2004; 427–430.

57. J.P. Laine, O. Gonnard, G. Charitat, L. Bertolini, and A. Peyre-Lavigne. Active pull-down protection for full substrate current isolation in smart power IC's. *Proceedings of the International Symposium on Power Semiconductor Devices (ISPSD)*, 2003; 273–276.

58. V. Khemka, V. Parthasarathy, R. Zhu, A. Bose, and T. Roggenbauer. Trade-off between high-side capability and substrate minority carrier injection in deep sub-micron smart power technologies. *Proceedings of the International Symposium on Power Semiconductor Devices (ISPSD)*, 2003; 241–244.

59. W. Horn. *On the Reverse-Current Problem in Integrated Smart Power Circuits*. Ph.D. Thesis, Technical University of Graz, Austria, April 2003.

11 Analog ESD Library and Documents

11.1 ANALOG DESIGN LIBRARY

In analog design technologies, the number of supported elements is significant [1, 2]. Analog and power technologies can support many power supply voltages leading to a large number of both passive and active elements. With the high number of power supply voltages, passive and active elements, the number of electrostatic discharge (ESD) devices and ESD circuits can also be significant [3–9]. With a large number of passive devices, active devices, and ESD networks, design errors can occur due to the following:

- Electrical overvoltage (EOV) and electrical overcurrent (EOC) of passive elements
- Electrical overstress (EOS) of active elements
- Misapplication and misuse of ESD circuits to analog circuits

For a bipolar-CMOS-DMOS (BCD) technology, the power supply voltage can range from low voltage CMOS to high voltage (HV) levels. For example, a 120 V BCD technology may support 1.8, 2.5, 5.0, 10, 16, 20, 25, 40, 60, and 120 V applications. For each one of these voltage application levels, there are both passive and active elements to support these voltages.

11.2 ANALOG DEVICE LIBRARY: PASSIVE ELEMENTS

Analog device libraries of supported devices can include resistors, capacitors, and inductors. In this section, analog passive elements will be reviewed.

ESD: Analog Circuits and Design, First Edition. Steven H. Voldman.
© 2015 John Wiley & Sons, Ltd. Published 2015 by John Wiley & Sons, Ltd.

11.2.1 Resistors

High-precision resistors are needed in analog applications, which contain high degree of matching and wide resistance ranges. Resistors in analog circuits include:

- Polysilicon resistors
- Metal resistors
- Silicon resistors

The resistors have different resistance tolerance values as well as different temperature coefficients of resistance (TCR) which influence analog application results.

For ESD applications, resistors are used as series resistor elements on signal pins. Resistors can also be used between power rails, between ground rails, and as RC-triggered MOSFET ESD power clamps. For these ESD applications, interdigitated layout practices are acceptable.

ESD robustness of resistor elements is a function of the cross-sectional area of the resistor, the material properties, and the layout and design [3–9]. The power-to-failure of resistor elements are a function of the cross-sectional area, thermal conductivity, heat capacity, melting temperature of the material, and the pulse width. For resistor elements, the ordering of the interconnect melting temperature, tungsten metallurgy has the highest melting temperature, copper, and then aluminum has the lowest.

For serpentine design styles, corner design can lead to a lower ESD failure levels. ESD failure due to current crowding on the corners is worse for 90° angles. ESD failure levels in serpentine design can be improved with chamfered corners.

For analog design, corner design issues can be eliminated using strap connections between segments of the resistor. In analog design, using contacts, vias, and metal connections, the accuracy and matching can be improved between the resistors. With the integration of strap connections, resistor elements can be interdigitated for matching within a circuit.

11.2.2 Capacitors

With a wide range of voltage applications, capacitor elements are needed to support the voltage conditions. Capacitor elements can consist of the following [5]:

- MOS capacitors
- Metal–insulator–metal (MIM) capacitors
- Metal–interlevel dielectric (ILD)–metal capacitors
- Vertical natural plate (VNP) or vertical parallel plate (VPP) capacitors

Capacitors are needed for ESD networks [3–5, 7–9]. Capacitor layout can be common centroid layout or interdigitated layout design but not required since matching is not critical for these circuits. For ESD power clamps, capacitors are used for RC-triggered ESD power clamps. ESD power clamps are typically used between the power and ground rails. RC-triggered power clamps can be used on low-dropout (LDO) regulators; buck, boost, and buck–boost regulators; and other voltage regulator outputs.

The capacitor elements have different levels of ESD robustness. The dielectric thickness influences the failure level of the capacitors.

11.2.3 Inductors

On-chip inductor designs are typically planar inductors formed from the metal wiring of the back end of line (BEOL) of a semiconductor chip. Semiconductor technology utilizes copper, aluminum, and tungsten metallurgies [3–5, 7–9]. For inductors, the melting temperatures of the metallurgy play a role in the ESD failure [5]. The on-chip spiral planar inductors have one of the following design styles:

- Square inductor
- Polygon inductor
- Octagonal inductor
- Circular inductor

These spiral inductors typically have an "underpass" at the center. The underpass connection is typically a lower-level metal design level below the spiral inductor coil.

Inductors can also be formed in intertwined pairs using multiple design levels. These can be used for circuits that require matched inductors or balun applications. Multilevel inductors take up less area than single-level spiral inductors.

11.3 ANALOG DEVICE LIBRARY: ACTIVE ELEMENTS

Analog device libraries require both active and passive elements [1, 2]. Analog design device libraries have a large variety of MOSFET and bipolar elements to fulfill a wide variety of power supply conditions. Analog libraries contain active elements of diodes and MOSFETs in a CMOS technology. For a bipolar-CMOS (BiCMOS) technology, analog libraries contain diodes, MOSFETs, and bipolar transistors.

Analog design libraries contain a variety of diode elements that can be used for circuits or ESD protection. Analog design libraries can consist of p+/n-well diodes, n+/substrate diodes, n-well-to-substrate diodes, Schottky diodes, and Zener diodes. Analog libraries can contain diodes with different isolation structures, from LOCOS, shallow trench isolation (STI), deep trench isolation, and polysilicon-bound gated diodes. Analog libraries can provide both isolated and nonisolated diodes. For noise reduction and higher isolation voltages, isolated diode elements can be utilized. Isolated diodes can be formed using triple well process technologies and technologies with buried layers.

Analog design libraries contain a variety of p-channel and n-channel MOSFET elements that can be used for circuits or ESD protection. Analog design libraries can consist of single-gate, dual-gate, and triple-gate oxide MOSFETs. Analog libraries can provide both isolated and nonisolated MOSFETs for noise reduction and higher isolation voltages. Isolated MOSFETs can be formed using triple well process technologies and technologies with buried layers. In LDMOS technology, LDMOS MOSFETs are added to the analog library and used for power device applications. Dual-gate and triple-gate oxide MOSFETs can be used for ESD protection for improved voltage tolerance.

Analog design libraries contain a variety of npn and pnp bipolar elements that can be used for analog circuits or ESD protection networks. Analog design libraries can consist of self-aligned and non-self-aligned bipolar transistors. Analog libraries can provide both isolated and nonisolated bipolar transistors for noise reduction and higher isolation voltages. Analog design libraries can consist of transistors of different emitter, base, and collector configurations, as well as symmetric and asymmetric design.

11.4 ANALOG DESIGN LIBRARY: REPOSITORY OF ANALOG CIRCUITS AND CORES

Analog design library systems can include basic circuit functions and cores that are supported by the design team. The advantage of the repository of analog circuit and cores is that the design has been verified for analog functional characteristics. In some corporations, the repository of analog circuits has been verified for ESD, latchup, and EOS.

11.4.1 Analog Design Library: Reuse Library

Analog design libraries can consist of a "reuse" library where circuit design teams place basic analog design functions into a repository for other teams to use. The reuse library may be a library which is not supported by the quality and reliability team. The problem with reuse library is that if they are not qualified, then EOS and ESD problems may be propagated across many designs. This is a concern with reused analog design with unverified ESD sensitivity. As a guideline, a good business practice is as follows:

- Test and quantify analog functional blocks for ESD and EOS sensitivity.

- Introduce a formal qualification and release process for reuse circuitry.

- Information notes and documentation on EOS and ESD test results on reuse functional blocks.

11.5 ESD DEVICE LIBRARY

With the high number of power supply voltages, the number of ESD devices and ESD circuits can also be significant. What is critical to avoid functional and ESD concerns is that the correct ESD device or element is associated with a compatible circuit. Misapplication or misuse can occur.

As discussed in the prior section, for a BCD technology, the power supply voltage can range from low voltage CMOS to HV levels. For example, a 120 V BCD technology may support 1.8, 2.5, 5.0, 10, 16, 20, 25, 40, 60, and 120 V applications. As a result, ESD elements are needed for each of these power supply values; for each power supply value, a signal pin and ESD power clamp may be required for each voltage level.

This leads to a very large ESD library of elements. With a large ESD library, circuit design teams will struggle to determine what is the proper ESD element or circuit for a given circuit application. There are solutions to address this issue and are as follows:

- ESD library is segmented by power supply voltage levels.

- ESD library is segmented into applications.

- ESD library is segmented by ESD type (e.g., ESD input circuit, ESD power clamps).

- ESD device or circuit cell name is clear and comprehensive.

- Reduction of ESD library elements size and number using hierarchical parameterized cells (PCells).

- ESD design manual section explaining supportive ESD elements.

- ESD "cookbook" to describe ESD cell naming convention and correspondence between circuit type and ESD cell to be used.

- ESD checking and verification methods to avoid misuse of ESD elements.

11.6 CADENCE-BASED PARAMETERIZED CELLS (PCELLS)

A design methodology is desirable that allows the optimization and tuning of ESD networks in an analog and mixed-signal environment. In an analog design environment, parameterized cells, also known as p-cells (e.g., or PCells), allow for a means to modify the size of the physical elements [5, 10–13]. Additionally, these designs must be able to integrate with guard ring technology requirements and under bond wire pads. Semiconductor devices, both active and passive elements, can be constructed from "primitive" p-cell. These primitive p-cell elements represent a single device element (e.g., resistor, capacitor, inductor, MOSFET, or bipolar element). These physical elements undergo full characterization, from which the released models are constructed. In analog ESD cosynthesis, it is desirable to be able to provide a methodology that allows the ability to vary the ESD network characteristics, whether size or topology, in order to evaluate the analog functional impacts. Given that the design methodology allows for both size and topology variations, characterization can be evaluated during the analog and ESD design phase.

An analog ESD design methodology which achieves this objective can be a method that forms ESD networks from these primitive p-cell device elements and converted to a higher-order parameterized cell [5, 10–13]. From the lowest-order device primitive parameterized cells, hierarchical parameterized cells can be constructed to form a library of ESD circuits and networks. By utilizing the primitive p-cell structures into higher-order networks, an ESD computer-aided design (CAD) strategy is developed to fulfill the objective as follows:

- Design flexibility
- DC characterization
- AC characterization
- Analog models
- Choice of ESD network type
- Compression of analog design library

Developing an ESD design system of hierarchical system of parameterized cells, higher-level ESD networks can be constructed without additional characterization. In this methodology, the lowest-order O[1] device p-cells can be DC and AC characterized. The basic device library is constructed of the devices, which are fully quantified; both passive and active elements are placed into DC pad sets for DC measurements. Additionally, the ESD testing can be completed on the base library of elements using wafer-level transmission line pulse (TLP) and human body model (HBM) testing. Note that the circuit characterization and ESD testing can be done on both the primitive O[1] p-cell elements or the O[n] hierarchical p-cell. In order to provide analog ESD design cosynthesis, it is desirable to have both the ability to vary physical size and circuit topology. In this method, it is possible with the generation of design layout, schematics, and symbology of both the primitive and the higher-order O[n] ESD p-cell networks.

11.6.1 ESD Hierarchical PCell Physical Layout Generation

In ESD design, it is desirable to change the physical layout of the ESD network. Hence, an RF-ESD design methodology is needed that can vary the physical layout of the design in physical size (e.g., area), form factor (e.g., length, width, and ratio), shape (e.g., rectangular, circular), and relative size of elements within a given circuit (e.g., be able to vary the size of the resistor, capacitor, or MOSFET independently within the common circuit topology) [5, 10–13]. For example, in some chip architectures, it is desirable to place multiple power clamps across a design, instead of a single element. It is also desirable to place different size ESD circuits for a common topology in a common chip. Hence, these features will be needed for integration and synthesis into a semiconductor chip architecture. In the formation of the hierarchical ESD p-cell elements, the generation of the ESD physical layout can be formed using the "graphical method" or the "code method" using SKILL code. In the "graphical method," the p-cells and physical shapes are placed in the design environment manually. The physical sizes of the O[1] p-cells are defined by its parameters. The physical dimensions that are desired to be modified are passed up to the higher-order p-cell design through "inheritance." The inherited parameters become free variables which allow the physical size changes of the O[1] elements. The parameters that allowed adjustment in each O[1] element are contained in the final

O[n] p-cell ESD network. Another methodology allows for the placement to be completed by software instead of manual placement of the physical elements and the electrical connections. SKILL code generates the schematic directly and forms placement of the elements.

In order for this design method to be successful, either the elements must be scalable or quantified. This can be verified through experimental testing of the ESD elements. Additionally, with the placement of a plurality of elements, new failure mechanisms must not occur. In our case, it was found that there is a range where linearity of ESD robustness versus structure size where this is true, and no new failure mechanisms were evident.

11.6.2 ESD Hierarchical PCell Schematic Generation

An analog ESD method that allows full circuit evaluation and the ability to change the physical size of the element through circuit simulation is desirable [10–13]. Additionally, it is desirable to provide changes in the circuit topology itself within the ESD network. The circuit topology can influence the trigger conditions, capacitance loading, linearity, and RF circuit stability. Hence, it is desirable to have ESD networks that can change the physical topology within a circuit design environment. Analog networks also have a wide range of application and power supply voltages; ESD circuits are required which can be modified to address the different application voltages. Hence, an ESD design system that allows for both change of circuit topology and structure size in an automated fashion is desirable. The circuit topology automation allows for the customer to autogenerate new ESD circuits and ESD power clamps without additional design work. In order for this method to be successful, the interconnect and associated wiring outside of the primitive elements must have a design environment that independently extracts the wiring, via, and interconnects. In our design environment, the wiring and via interconnect are autoextracted, which then eliminates the need for evaluation of electrical characteristics for every model element and every ESD structure.

11.6.3 ESD Design with Hierarchical Parameterized Cells

In semiconductor chip design, there are fundamental ESD functional blocks required in the semiconductor chip synthesis and floor planning. For ESD design, the minimum set of classes of hierarchical parameterized cells needed to support an ESD design system involved are as follows [10–13]:

• ESD input networks

• ESD rail-to-rail networks

• ESD power clamps

In a mixed-signal ESD design system, the ESD design system should contain a family of these elements in each category which satisfy signal types (e.g., digital, analog, and power), analog parameters (e.g., noise, linearity, stability), and spatial placement. For analog input circuits, the loading effects of ESD input and impedance influence are critical for performance evaluation.

For ESD rail-to-rail networks, the issue is stability and noise coupling is important in mixed-signal environments. Noise is a concern in digital networks; peripheral and core circuitries are isolated when the peripheral circuit noise is significant and the interior core logic networks are sensitive to noise disruption. Noise is a large concern in semiconductor chips with both digital and analog function on a common substrate. In mixed-signal applications, functional circuit blocks are separated to minimize noise concerns. Digital noise affects both the analog and DC circuitry impacting the noise figure (NF). Designers need the ability to estimate the noise and stability of the circuit in the presence of multiple circuits and ESD networks. To eliminate noise, digital circuit blocks are separated from the analog circuit blocks without a common ground or power bus. The introduction of the ESD elements between the grounds can address the ESD concerns but increases the noise and stability implications. As a result, the cosynthesis of the ESD and noise concerns needs to be flexible to address both issues. And, for ESD power clamps, the ESD networks can influence noise, stability, and leakage.

11.6.4 Hierarchical PCell Graphical Method

For construction of the p-cell, there are different methods of p-cell definition within the Cadence™ environment. This methodology is referred to as the "graphical" technique [10–13]. The command structure for p-cell definition involves Stretch, Conditional Inclusion, Repetition, Parameterized Shapes, Repeat along Shape, Reference Point, Inherited Parameters, Parameterized Layer, Parameterized Label, Parameterized Property, Parameters, and Compile. The Stretch function allows Stretch in X, Stretch in Y, Qualify, and Modify. The Repetition function allows for Repeat in X, Repeat in Y, and Repeat in X and Y.

Stretch commands require an algorithm to define the design "expression for stretch." For this p-cell, the "expression for stretch" is defined as {{pitch * num_stripes_up} - pitch} where "pitch" is the width of the upward diode periodicity and the "num_stripes_up" is an inherited parameter contained in the higher-order p-cell passed from the lower p-cell to address the number of fingers of the diode between the input pad and the V_{DD} power supply. Likewise for the downward diode, a second "expression for stretch" is defined for the second p-cell diode element stretch line. The expression for stretch is defined as "{{pitch*num_stripes_down} - pitch}" for the second stretch line in the y-direction. For the first stretch line, the direction of stretch is "up"; for the second stretch line, the direction of stretch is "down." For the stretch of the diode p-cells, and the busses, a stretch line exists in the x-direction. For the stretch in x, an "expression of stretch" is defined as {{a * num_segments_up} + b} where a and b are constants. The stretch direction is chosen to the right.

For the P-cell parameter summary, a typical output has the form as shown below:

Parameters defined in this parameterized cells

```
num_segments_up num_segments_down
num_stripes_up num_stripes_down
Stretch
```

```
Stretch Type: Vertical
Name of Expression of Stretch: 2.52*num_stripes_down - 2.52
Stretch Direction: down
Stretch
Stretch Type: Vertical
Name of Expression of Stretch: 2.52*num_stripes_up - 2.52
Stretch Direction: up
Stretch
Stretch Type: Horizontal
Name of Expression of Stretch:
7.6*num_segments_up + 4.2
Stretch Direction: up
```

Inherited Parameters

```
Number of instances with parameter inheritance: 2
Instance Name: I3
Inherited Parameter: Name: num_X
Inherited Parameter: Value: num_segments_up
Inherited Parameter: Type: integer
Inherited Parameter: Default: 1
Inherited Parameter: Name: num_PD
Inherited Parameter: Value: num_stripes_up
Inherited Parameter: Type: integer
Inherited Parameter: Default: 1
Inherited Parameter: Name: num_X
Inherited Parameter: Value: num_segments_down
Inherited Parameter: Type: integer
Inherited Parameter: Default: 1
Inherited Parameter: Name: num_PD
Inherited Parameter: Value: num_stripes_down
Inherited Parameter: Type: integer
Inherited Parameter: Default: 1
```

As part of an analog ESD CAD design system, a hierarchical O[3] parameterized cell is designed which forms bidirectional O[2] series diode strings which can vary the number of series diode elements and the physical width of each diode element. For example, a design may use four diodes in one direction and two in the other direction between the ground rails. The automated ESD design system has the ability to adjust the design size and the number of elements. In digital circuits, the design decision is typically decided based on the digital DC voltage separation required between the grounds; in high-speed digital and analog circuits, the design issue is the capacitive coupling at high frequency. As more elements are added, capacitive coupling is reduced. In our ESD design system, the interconnects and wires automatically stretch and scale with the structure size. Algorithms are developed which autogenerate the interconnects based on the number of diodes "up" versus diodes "down." As elements are added, both the graphical layout and physical schematics introduce the elements maintaining the electrical interconnects and pin connection.

11.6.5 Hierarchical PCell Schematic Method

A powerful feature of an automated ESD design system is the ability to autogenerate ESD networks from both the "graphical method" and the "schematic method." Typically, ESD designers start from the graphical layout of the physical design, and circuit designers start from the schematic layout to evaluate the performance objectives. To cosynthesize the performance and the ESD objectives, it is important to be able to have different modes of integration and starting points in the design methodology.

To achieve autogeneration of ESD circuits, a design flow has been developed (Figure 11.1). The flow is based on the development of p-cells for both the schematic and layout cells. The p-cells are hierarchical, built from device O[1] primitives which have been characterized with defined models. Without the need for additional RF characterization, the design kit development cycle is compressed. Autogeneration also allows for design rule checking (DRC) correct layouts and layout-versus-schematic (LVS) correct circuits.

In the ESD CAD design system, the schematic p-cell is generated by the input variables to account for the inherited parameters of input values. A problem with schematic autogeneration is the circuit simulation phase. The circuit may be placed as a subcircuit; however, specter simulation will only allow a single definition of a subcircuit. This prevents the reuse of the schematic p-cell in any other configuration. To retain the ESD circuit variability, a design flow has been built around the schematic p-cell.

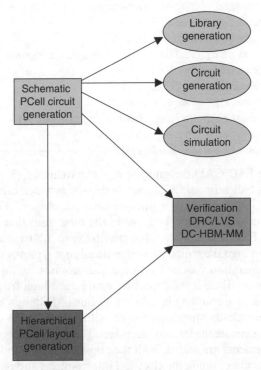

Figure 11.1 CAD ESD design flow.

From the schematic methodology, four different modes of implementation were addressed:

- Creation of the ESD element
- Creation and placement of an ESD element
- Placement of an existing ESD element
- Placement of an ESD schematic

In an automated design environment, Figure 11.3 shows a representation of the different methodologies. As an example of the schematic methodology, from the schematic editing screen, the user invokes AMS utils → ESD. From the ESD pull-down, four functions are defined: ESD → Create an ESD element, ESD → Create and Place an ESD element, ESD → Place an existing ESD element, and ESD → Place an ESD schematic.

In our ESD CAD design system, the schematic PCell is generated by the input variables to account for the inherited parameters of input values. A problem with schematic autogeneration is the circuit simulation phase. The circuit may be placed as a subcircuit; however, simulation will only allow a single definition of a subcircuit. This prevents the reuse of the schematic p-cell in any other configuration. To retain the ESD circuit variability, a design flow has been built around the schematic p-cell.

In one method, the designer is allowed the capability of building an ESD library with the creation of ESD cells. The designer will select the option to "Create an ESD element." Figure 11.2 shows an example of where the "Create an ESD element" function initiates

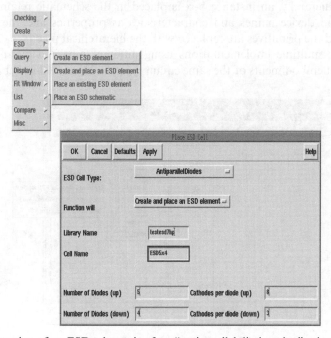

Figure 11.2 Creation of an ESD schematic of an "antiparallel diode string" using the schematic method.

creation of an ESD schematic for a parameterized cell of a back-to-back diode string known as "AntiparallelDiodeString." To generate the electrical schematic, the ESD design system requests the "number of diodes up" and the "number of diodes down"; this determines the number of diodes in the string that are used between digital V_{SS} and analog V_{SS} (or RF V_{SS}) for grounds. For power supply rails, the "AntiparallelDiodeString" is used between digital V_{DD} and analog V_{DD} (or RF V_{DD}). The design system also requests the number of cathode fingers in the diode structures for the "up" string and "down" string. The input parameters are passed into a procedure which will build an ESD cell with the schematic p-cell built according to the input parameters and placed in the designated ESD cell. An instance of the ESD layout p-cell will also be placed in the designated ESD cell. This allows for the automated building of an ESD library, creating a schematic, layout, and symbol of the circuit based on the input parameters. This symbol may be placed in the circuit by selecting the "Place an ESD circuit" option.

The second method allows for the autogeneration of the schematic ESD circuit to be placed directly into the design. This procedure available with the "Place an ESD schematic" option will allow the designer to autogenerate the circuit and place it in the schematic. Since these cells are hierarchical, the primitive devices and autowiring are placed by creating an instance of the schematic p-cell and then flattening the element. The instance must be flattened to avoid redefinition of subcircuits. Figure 11.3 shows the generation of the back-to-back antiparallel diode string.

The problem arises during the layout phase of the design. In the schematic due to the flattening, the hierarchy has been removed, and only primitive elements remain. During design implementation, the primitives will be placed, and the hierarchy will be lost. To maintain the hierarchy, an instance box is placed in the schematic retaining the input parameters and device names and characteristics as properties and the elements are recognized and the primitives are replaced with the hierarchical p-cell.

To produce multiple implementations using different inherited parameter variable input, different embodiments of the same circuit type can be created in our methodology.

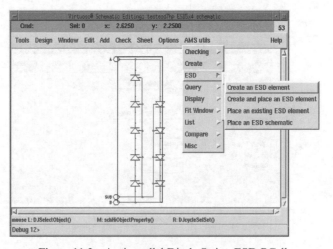

Figure 11.3   Antiparallel Diode String ESD PCell.

In this process, the schematic is renamed to be able to produce multiple implementations in a common chip or design; the renaming process allows for the design system to distinguish multiple cell views to be present in a common design.

When the inherited parameters are defined, the circuit schematic is generated according to the selected variables. Substrate, ground, and pin connections are established for the system to identify the connectivity of the circuit. The design system can also autogenerate the layout from the electrical schematic which will appear as equivalent to the previously discussed graphical implementation.

The physical layout of the ESD circuits is implemented with p-cells using existing primitives in the reference library. The circuit topology is formed within the p-cell including wiring such that all parasitics may be accounted for in preproduction test site construction.

In a Bipolar, BiCMOS, and BCD technology, ESD networks can be constructed from these primitive p-cell elements of CMOS, bipolar, and LDMOS elements. Hierarchical parameterized cells can be formed from bipolar-only primitive p-cells, CMOS-only primitive p-cells, LDMOS p-cells, or primitive elements. Additionally, new primitive elements using hybrid levels can be used for the primitive cell design. From the lowest-order O[1] primitive parameterized cells, hierarchical parameterized cells can be constructed to form a library of ESD networks. In the list of primitive p-cell elements, the hybrid primitive elements can provide advantages in ideality, capacitance, substrate injection, noise isolation, latchup robustness, and ESD robustness.

In this mixed-signal ESD design system methodology, there are significant advantages and limitations. This methodology developed has been adopted in a foundry environment and has demonstrated significant unanticipated advantages in an analog and mixed-signal design team environment:

- A significant improvement in design and release productivity is evident from implementation of the hierarchical parameterized cell ESD library.

- The designs are completed at the test site phase; this allows for direct implementation into the design kit release/verification process at the test site phase of the development cycle.

- The hierarchical parameterized ESD designs do not need unique DC and AC characterization since the designs contain all AC characterized elements; this provides no additional characterization workload and will be updated with all design releases.

- The customers do not continue to request alternate size structures of different form factors after the initial release.

- As the design system matures, the number of inherited parameters can be increased to allow increased customer flexibility to address area, form factor, or other issues.

- The modular nature of the hierarchical ESD designs allows reuse and flexibility. The introduction of new ESD networks is possible utilizing the existing implementation and modifying the hierarchy for the modification.

11.7 ANALOG ESD DOCUMENTS

ESD documents are important for analog and mixed-signal design. In the following sections, technology design manual, cookbooks, and checklists for ESD and EOS are discussed. Figure 11.4 shows an example of an EOS control program documents that can be applied to an analog technology.

11.7.1 ESD Technology Design Manual Section

ESD technology design manual sections are important for defining the usage of ESD elements and the supported networks. ESD technology design manuals are typically contained within the technology design manual. Technology design manuals contain all the design rules for the specific technology.

Historically, in early ESD development, the ESD technology design manual section was small and not comprehensive. In the 1980s, the ESD design manual section contained one or two ESD elements, a few recommendations, and no ESD DRC rules. The analog ESD design guidelines utilized the same solution as the digital ESD guidelines. Today, ESD technology design manual sections have been significantly expanded.

ESD technology design manual sections can include the following:

- ESD required specifications

- ESD supported standards

- ESD supported designs

- ESD design rules

- ESD design recommendations

- ESD guard ring rules

- ESD layout design practices

- Do's and Don'ts

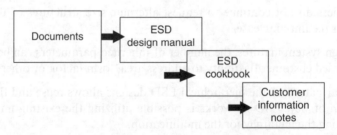

Figure 11.4 ESD control program documents.

11.7.1.1 ESD Required Specifications

Design manuals include the ESD required specifications for the objectives of the technology and applications. The required ESD, EOS, and latchup testing corporate objectives can be included in the design manual. Examples of required ESD test requirements and specification level are as follows:

- HBM—2000 V ESD specification level

- Charged device model (CDM)—1000 V ESD specification level

11.7.1.2 ESD Supported Standards

Design manuals include the ESD required supported standards for the technology and applications. The required ESD, EOS, and latchup testing corporate requirements can be included in the design manual. Examples of ESD test-supported requirements are as follows [14–23]:

- HBM: JESD22-A114E

- MM: EIA/JESD22-A115-A

- CDM: JESD22-C101C

- IEC 61000-4-2

- IEC 61000-4-5

Existing ESD standards include the following [14–23]:

11.7.1.2.1 Human Body Model (HBM)

ANSI/ESD ESD-STM 5.1-2007. ESD Association Standard Test Method for the Protection of Electrostatic Discharge Sensitive Items—Electrostatic Discharge Sensitivity Testing—Human Body Model (HBM) Testing—Component Level. Standard Test Method (STM) document, 2007.

11.7.1.2.2 Machine Model (MM)

ANSI/ESD ESD-STM 5.2-1999. ESD Association Standard Test Method for the Protection of Electrostatic Discharge Sensitive Items—Electrostatic Discharge Sensitivity Testing—Machine Model (MM) Testing—Component Level. Standard Test Method (STM) document, 1999.

11.7.1.2.3 Charged Device Model (CDM)

ANSI/ESD ESD-STM 5.3.1-1999. ESD Association Standard Test Method for the Protection of Electrostatic Discharge Sensitive Items—Electrostatic Discharge Sensitivity Testing—Charged Device Model (CDM) Testing—Component Level. Standard Test Method (STM) document, 1999.

JEDEC. JESD22-C101-A. A Field-Induced Charged Device Model Test Method for Electrostatic Discharge-Withstand Thresholds of Microelectronic Components, 2000.

11.7.1.2.4 IEC 61000-4-2

IEC. IEC 61000-4-2. Electromagnetic compatibility (EMC)—Part 4-2: Testing and measurement techniques—Electrostatic discharge immunity test. *IEC International Standard*, 2007.

11.7.1.2.5 Human Metal Model (HMM)

ESD Association. ESD-SP5.6-2008. ESD Association Standard Practice for the Protection of Electrostatic Discharge Sensitive Items—Electrostatic Discharge Sensitivity Testing—Human Metal Model (HMM) Testing Component Level. Standard Practice (SP) document, 2008.

11.7.1.2.6 Transmission Line Pulse (TLP)

ANSI/ESD Association. ESD-SP5.5.1-2004. ESD Association Standard Practice for the Protection of Electrostatic Discharge Sensitive Items—Electrostatic Discharge Sensitivity Testing—Transmission Line Pulse (TLP) Testing Component Level. Standard Practice (SP) document, 2004.

ANSI/ESD Association. ESD-STM 5.5.1-2008. ESD Association Standard Test Method for the Protection of Electrostatic Discharge Sensitive Items—Electrostatic Discharge Sensitivity Testing—Transmission Line Pulse (TLP) Testing Component Level. Standard Test Method (STM) document, 2008.

11.7.1.2.7 Very Fast Transmission Line Pulse (VF-TLP)

ANSI/ESD Association. ESD-SP5.5.2-2007. ESD Association Standard Practice for the Protection of Electrostatic Discharge Sensitive Items—Electrostatic Discharge Sensitivity Testing—Very Fast Transmission Line Pulse (VF-TLP) Testing Component Level. Standard Practice (SP) document, 2007.

ESD Association. ESD-STM 5.5.2-2009. ESD Association Standard Test Method for the Protection of Electrostatic Discharge Sensitive Items—Electrostatic Discharge Sensitivity Testing—Very Fast Transmission Line Pulse (VF-TLP) Testing Component Level. Standard Test Method (STM) document, 2009.

11.7.1.3 ESD Supported Designs

ESD supported designs can be placed in the design manual section. To avoid usage of unsupported design, it is important to have the layout and design teams to be aware of the designs that are supported. This can be addressed by identifying the designs in a supported Cadence library.

11.7.1.4 ESD Design Rules

ESD design rules should be contained in the technology design manual. Design rules are the rules the design system "check and verify" through the design system. The ESD design rules can include the following:

- ESD signal pin network requirements (e.g., type, width, size)

- ESD power clamp network requirements (e.g., type, width, size)

- Signal pin wire interconnect width requirements

- Power rail width and/or resistance rule

- Spatial placement of ESD power clamp relative to signal pins

- Domain-to-domain ESD network requirements

- Domain-to-domain signal wire requirements

- I/O guard ring rules

- I/O-to-I/O rules

- I/O-to-core domain rules

- Core-to-core rules

- Internal latchup rules

- External latchup rules

11.7.1.5 ESD Design Recommendations
ESD design recommendations should be contained in the technology design manual. A design recommendation is not required to be a design system rule that requires checking and verification.

11.7.1.6 ESD Guard Ring Rules
ESD guard ring rules can include the requirements of the ESD networks themselves, and those relative to adjacent structures. ESD guard ring rules can include the following:

- ESD signal pin element guard rings

- ESD power clamp guard ring

- ESD digital-to-analog rail-to-rail guard ring rules

- ESD signal pin-to-I/O guard ring rules

- ESD power clamp-to-I/O guard ring

- ESD signal pin-to-analog core domain rules

11.7.1.7 ESD Layout Design Practices
ESD layout requirements and design practices can be included in the design manual. In this section, interdigitated layout and common centroid practices can be integrated into the ESD layout and cosynthesis with the analog circuitry.

11.7.1.8 Do's and Don'ts
ESD layout practices are at times demonstrated as drawings to allow the layout and design teams to practice good layout practices. Some technology design manuals demonstrate these as good versus bad design practices (e.g., "Do's and Don'ts" practices).

11.8 ESD COOKBOOK

In early development of analog corporations, in the 1990s, it was realized that with the complexity and breadth of the number of technologies supported, additional documents were needed for circuit design teams to address ESD issues [8, 9]. A separate document evolved which designers referred to as an "ESD cookbook." ESD cookbooks in analog corporations provided ESD circuits, recommendations, and rules. Each section of the ESD cookbook was separated for each technology type and generation (Figure 11.5). The ESD cookbook supported bipolar, MOSFET, and power technologies, where the device set and rules were significantly different.

An ESD cookbook will have specific goals for the circuit design community to guarantee correct implementation [8, 9]. Some of the specific goals are as follows:

- **Choosing the correct technology:** The first goal is to determine the correct technology supported ESD network for the corresponding circuit. Tables will be formed of digital, analog, and power network applications, which identify common functional blocks. The tables will contain circuit type, application voltage, the ESD element to be used to protect the functional block, and ESD cell name.

- **ESD circuit description:** The second goal is to have the ESD circuits described with the circuit schematic, cell name, and symbol cell view for quick reference.

- **ESD data of supported ESD networks:** The third goal is to have ESD data (HBM, MM, and TLP) for the supported networks as a function of the structure size.

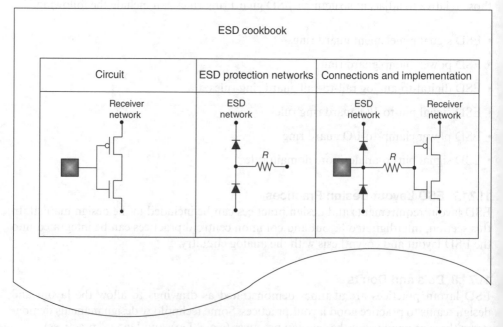

Figure 11.5   ESD analog cookbook.

- **ESD placement:** The fourth goal is to assist the circuit designer in the rules for placement of the ESD signal pin networks and ESD power clamp networks.

- **Interconnects and connecting:** The fifth goal is to assist the circuit designer in proper wiring of the ESD network in the chip design synthesis. This is to include wire width, distribution, and guidelines from proper integration of the ESD circuits with the full chip design synthesis.

- **Repository of information:** The sixth goal is to have a repository for customer information notes (CIN), guidelines, and concerns with the various networks.

The ESD cookbook must provide guidance to the circuit design teams. In addition, the document is to provide guidance for the following [9]:

- **Usage of appropriate ESD signal pin and power networks for circuit and product application:** Choose appropriate size based on ESD specifications type and magnitude of product requirements.

- **Identify the appropriate supported ESD network for the given voltage requirements.**

- **Placement the peripheral I/O circuitry:** Provide guard rings based on supported ESD and latchup design rules.

- **Establish power and ground bus width and via number for adequate width to support I, O, and I/O circuits and ESD networks:** Provide wire width and via number based on wire width and via number design rules.

- **Placement of the ESD signal pin networks:** Provide adequate metal and via width for signal pad to ESD network based on wire width and via number design rules.

- **Placement of the ESD power rail networks (e.g., V_{DD} to V_{SS}, V_{SS} to V_{SS}):** Verify worst-case resistance between the farthest signal pin and the ESD power rail networks (e.g., ESD transient clamps and ground to ground).

- **Verify bidirectional current paths between signal pins and power rails:** Bidirectional current paths must exist between pin-to-rail and rail-to-rail of common or separated function domains.

- **Checklist:** Complete ESD checklist for the product release team from the design conception to the final product release.

- **Product release:** Test the ESD product according to the corporate ESD test requirements and the supported ESD standards.

11.9 ELECTRICAL OVERSTRESS (EOS) DOCUMENTS

For EOS, it is important to include representation for various members of the design team in the release and sign-off process [9]. In the EOS design release, there are members of the team from application engineering, product definition, device design, circuit design, packaging, and reliability engineering.

11.9.1 EOS Design Release Process

An EOS design release process should contain some of the following items:

- Package EOS requirement
- Bond wire EOS requirement
- On-chip EOS protection
- On-chip ESD protection
- On-chip power rail requirements for EOS and ESD
- On-chip interconnect width requirements for EOS and ESD
- EOS robustness of technology elements
- Product—application EOS and ESD compatibility
- Product—technology EOS and ESD compatibility
- EOS DRC results
- EOS LVS results
- EOS electrical rule check (ERC) results

By establishing a good semiconductor chip release process, the risks for EOS can be reduced for semiconductor manufacturing release. Figure 11.6 shows an example of an EOS control program that can be applied to an analog technology.

11.9.2 Electrical Overstress (EOS) Cookbook

In the product definition, design integration, and planning stage, many corporations have developed "cookbook" documents for circuit and printed circuit board (PCB) design teams [8, 9]. An EOS cookbook can be used by the entire team to avoid design implementation errors and insure product success.

By establishing a good product release process, the risks for EOS can be reduced for semiconductor manufacturing release and EOS robustness in the field. Whereas an EOS or ESD design manual may refer to the different EOS and ESD design, an "EOS cookbook" will provide information of correspondence between circuit type and the protection elements. An EOS cookbook should contain some of the following items:

- Guideline on how to use the document
- EOS specification requirements for qualification
- EOS and ESD models

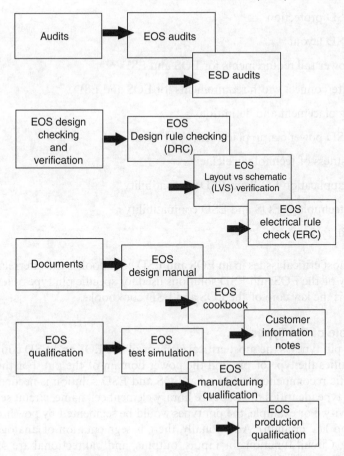

Figure 11.6 EOS control program.

- EOS and ESD specification references
- Off-chip EOS protection solutions library
- Advantages and disadvantages of different EOS protection solutions
- Off-chip EOS protection layout
- Off-chip EOS guidelines
- PCB EOS protection solutions
- Package EOS requirement
- Bond wire EOS requirement
- On-chip EOS protection

- On-chip ESD protection

- On-chip ESD layout

- On-chip power rail requirements for EOS and ESD

- On-chip interconnect width requirements for EOS and ESD

- Guard ring placement and definition

- On-chip ESD power clamp placement

- EOS robustness of technology elements

- Product—application EOS and ESD compatibility

- Product—technology EOS and ESD compatibility

- Table of pin types

One of the most critical issues in an EOS and ESD cookbook is the interrelationship and compatibility of the EOS and ESD solutions used for specific pin types. Hence, the table of pin types is the key core of the EOS and ESD cookbooks.

11.9.2.1 Table of Pin Types

The table of pin types is the most critical element of the EOS and ESD cookbooks. This section identifies the type of pin and the power domain of the pin. For the specific pin type, a specific recommendation of what EOS and ESD solution is needed. In this section, the pin type identifies the Cadence library element cell name, circuit schematic, and the connectivity. For example, the pin types would be segmented by power domain (e.g., 2.5, 5.0 V, and HV domain). Additionally, there is segmentation of analog, digital, and HV circuitry. Circuit functions of inputs, outputs, and bidirectional are segmented for the different ESD and EOS solutions. In the end, clear guidelines are set for the circuit design teams so that the correct EOS and ESD circuit is used for the specific circuit to avoid mismatch between the product application voltage tolerance of the elements and proper "turn-on" of the different EOS and ESD elements for the specific circuit (Figure 11.7).

The pin type table for the different voltage domains can include the following [9]:

- V_{DD} power input

- V_{DD} power input/output

- Digital input

- Digital output

- Digital open-drain output

- Bidirectional I/O

- Analog input

- Analog output

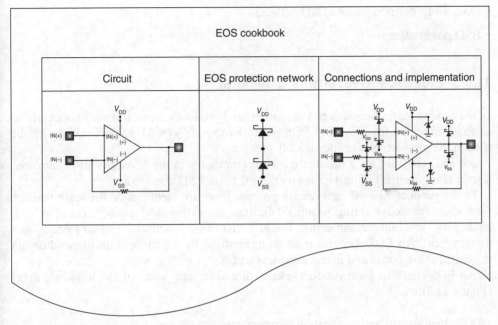

Figure 11.7   EOS cookbook.

The pin type table and corresponding ESD network for HV domains can include the following:

- VIN or V_{DD} power input
- HV digital input
- HV analog output
- HV open-drain output
- HV phase output
- HV phase input
- Analog output

The ESD design networks can include isolated and nonisolated designs. For power technologies, the ESD designs can include the following:

- Nonisolated ESD networks
- Isolated ESD networks (5, 10, 40, 60, 80, and 120 V)

Isolated and nonisolated designs are required for the following:

- Primary stage signal pin ESD
- Secondary CDM signal pin ESD

- Domain-to-domain ground ESD networks

- ESD power clamps

11.9.3 Electrical Overstress Checklist

In the product definition, design integration, and product release, it is very important to establish checklist for providing EOS product robustness [8, 9]. EOS checklists can be used in the product release and sign-off process.

An ESD checklist for a product release is commonly in the development of semiconductor components. Figure 11.8 is an example of an ESD checklist.

In the product sign-off and release process, there are members of the team that are from application engineering, product definition, marketing, device design, circuit design, packaging, reliability engineering, quality engineers, functional test engineers, and management. An EOS checklist must be understood by the entire team to avoid design implementation errors and insure product success.

An EOS checklist for a product release should contain some of the following items (Figure 11.9):

- EOS environment and application requirements

- ESD environment and application requirements

- Product specifications and EOS implications

- Off-chip EOS protection solutions

- PCB EOS protection solutions

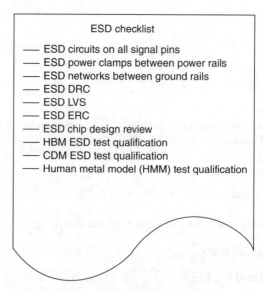

Figure 11.8   ESD checklist.

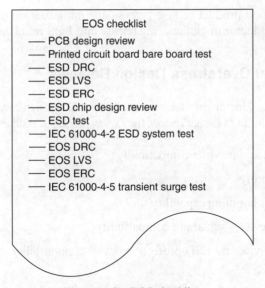

Figure 11.9 EOS checklist.

- PCB electrical characteristics
- Package EOS requirement
- Bond wire EOS requirement
- On-chip EOS protection
- On-chip ESD protection
- On-chip power rail requirements for EOS and ESD
- On-chip interconnect width requirements for EOS and ESD
- EOS robustness of technology elements
- Product—application EOS and ESD compatibility
- Product—technology EOS and ESD compatibility
- DRC results
- LVS results
- ERC results
- Functional test results
- ESD test results
- EOS test results
- System assembly procedures
- Manufacturing audit results
- Sign-off release of all organizations (e.g., technology, product definition, quality)

By establishing a good product release process, the minimization of EOS losses can be reduced for semiconductor manufacturing release and EOS robustness in the field.

11.9.4 Electrical Overstress Design Reviews

As part of the semiconductor chip release process, an EOS design review should be part of the process [8, 9]. In an EOS design review, the review process should evaluate the following:

- EOS environment and product compatibility

- Package compatibility

- Wire bond current handling capability

- EOS protection device—signal pin compatibility

- EOS protection device and ESD protection device compatibility

- EOS DRC results

- EOS LVS results

- EOS ERC results

- Electromigration maximum current density compatibility

In the EOS design review process, visual inspection of traces on the PCB and interconnect wiring levels to improve the maximum current carrying conditions is valuable to build EOS robust products.

11.10 CLOSING COMMENTS AND SUMMARY

In this chapter, ESD and EOS libraries and documents for an analog or mixed-signal technology were discussed. The discussion includes a plethora of items, from analog libraries, ESD library elements, Cadence-based parameterized cells, and Cadence-based hierarchical ESD designs. ESD and EOS documents for technology design manual, cookbooks, checklists, and design release processes are discussed.

In Chapter 12, ESD and latchup checking and verification methods are discussed. ESD DRCs for implementing ESD networks into the design are reviewed. Latchup DRCs for both internal and external latchup are highlighted as part of the design implementation. A key issue is the checking and verification of analog-to-digital cross-domain signal lines.

REFERENCES

1. V. Vashchenko and A. Shibkov. *ESD Design for Analog Circuits*. New York: Springer, 2010.
2. H. Kunz, G. Boselli, J. Brodsky, M. Hambardzumyan, and R. Eatmon. An automated ESD verification tool for analog design. *Proceedings of the Electrical Overstress/Electrostatic Discharge (EOS/ESD) Symposium*, 2010; 103–110.

3. S. Voldman. *ESD: Physics and Devices*. Chichester, UK: John Wiley & Sons, Ltd, 2004.

4. S. Voldman. *ESD: Circuits and Devices*. Chichester, UK: John Wiley & Sons, Ltd, 2005.

5. S. Voldman. *ESD: RF Circuits and Technology*. Chichester, UK: John Wiley & Sons, Ltd, 2006.

6. S. Voldman. *Latchup*. Chichester, UK: John Wiley & Sons, Ltd, 2007.

7. S. Voldman. *ESD: Failure Mechanisms and Models*. Chichester, UK: John Wiley & Sons, Ltd, 2009.

8. S. Voldman. *ESD Basics: From Semiconductor Manufacturing to Product Use*. Chichester, UK: John Wiley & Sons, Ltd, 2012.

9. S. Voldman. *Electrical Overstress (EOS): Devices, Circuits, and Systems*. Chichester, UK: John Wiley & Sons, Ltd, 2013.

10. S. Voldman, S. Strang, and D. Jordan. An automated electrostatic discharge computer-aided design system with the incorporation of hierarchical parameterized cells in BiCMOS analog and RF technology for mixed signal applications. *Proceedings of the Electrical Overstress/ Electrostatic Discharge (EOS/ESD) Symposium*, October 2002; 296–305.

11. S. Voldman, S. Strang, and D. Jordan. A design system for auto-generation of ESD circuits. *Proceedings of the International Cadence Users Group*, September 2002.

12. S. Voldman. Automated hierarchical parameterized ESD network design and checking system. U.S. Patent No. 5,704,179, March 9, 2004.

13. D. Collins, D. Jordan, S. Strang, and S. Voldman. ESD design, verification, and checking system and method of use. U.S. Patent Application 20,050,102,644, May 12, 2005.

14. ANSI/ESD ESD-STM 5.1-2007. *ESD Association Standard Test Method for the Protection of Electrostatic Discharge Sensitive Items—Electrostatic Discharge Sensitivity Testing—Human Body Model (HBM) Testing—Component Level*. Standard Test Method (STM) document, 2007.

15. ANSI/ESD ESD-STM 5.2-1999. *ESD Association Standard Test Method for the Protection of Electrostatic Discharge Sensitive Items—Electrostatic Discharge Sensitivity Testing— Machine Model (MM) Testing—Component Level*. Standard Test Method (STM) document, 1999.

16. ANSI/ESD ESD-STM 5.3.1-1999. *ESD Association Standard Test Method for the Protection of Electrostatic Discharge Sensitive Items—Electrostatic Discharge Sensitivity Testing— Charged Device Model (CDM) Testing—Component Level*. Standard Test Method (STM) document, 1999.

17. JEDEC. JESD22-C101-A. *A Field-Induced Charged Device Model Test Method for Electrostatic Discharge-Withstand Thresholds of Microelectronic Components*, 2000.

18. IEC. IEC 61000-4-2. Electromagnetic compatibility (EMC)—Part 4-2: Testing and measurement techniques—Electrostatic discharge immunity test. *IEC International Standard*, 2007.

19. ESD Association. ESD-SP5.6-2008. *ESD Association Standard Practice for the Protection of Electrostatic Discharge Sensitive Items—Electrostatic Discharge Sensitivity Testing— Human Metal Model (HMM) Testing Component Level*. Standard Practice (SP) document, 2008.

20. ANSI/ESD Association. ESD-SP5.5.1-2004. *ESD Association Standard Practice for the Protection of Electrostatic Discharge Sensitive Items—Electrostatic Discharge Sensitivity Testing—Transmission Line Pulse (TLP) Testing Component Level*. Standard Practice (SP) document, 2004.

21. ANSI/ESD Association. ESD-STM 5.5.1-2008. *ESD Association Standard Test Method for the Protection of Electrostatic Discharge Sensitive Items—Electrostatic Discharge Sensitivity Testing—Transmission Line Pulse (TLP) Testing Component Level*. Standard Test Method (STM) document, 2008.

22. ANSI/ESD Association. ESD-SP5.5.2-2007. *ESD Association Standard Practice for the Protection of Electrostatic Discharge Sensitive Items—Electrostatic Discharge Sensitivity*

Testing—Very Fast Transmission Line Pulse (VF-TLP) Testing Component Level. Standard Practice (SP) document, 2007.

23. ESD Association. ESD-STM 5.5.2-2009. *ESD Association Standard Test Method for the Protection of Electrostatic Discharge Sensitive Items—Electrostatic Discharge Sensitivity Testing—Very Fast Transmission Line Pulse (VF-TLP) Testing Component Level.* Standard Test Method (STM) document, 2009.

12 Analog ESD and Latchup Design Rule Checking and Verification

In this chapter, the focus is on electronic design automation (EDA) associated with electrical overstress (EOS) design. This continues to be a growing issue, as new concepts and techniques are developed. In this chapter, electrostatic discharge (ESD) concepts will also be highlighted due to the similarity of concepts for both EOS and ESD EDA. Latchup electronic design automation (EDA) techniques relevant to EOS issues will also be shown.

12.1 ELECTRONIC DESIGN AUTOMATION

EDA is a software tool specifically for the design of electronic systems such as single components, integrated circuits, and printed circuit boards (PCBs). EDA tools can be applied to ESD [1–28], EOS [25], latchup [11, 29–54], and other electromagnetic compatibility (EMC) issues. In this chapter, the focus will be on how EDA and computer-aided design (CAD) can be utilized to provide more robust electronic systems addressing whole-chip analysis and cross-domain issues in mixed-signal (MS) system-on-chip (SOC) applications [23, 55–69]. Figure 12.1 is an example of the checking and verification needs today in an EOS environment.

12.2 ELECTRICAL OVERSTRESS (EOS) AND ESD DESIGN RULE CHECKING

EOS and ESD design rules will be discussed in this section [1–3]. As new problems and issues arise in integrated circuits and SOC applications, these rules continue to be added. In this chapter, we will touch on a few examples of how to apply these to EOS, ESD, and latchup.

ESD: Analog Circuits and Design, First Edition. Steven H. Voldman.
© 2015 John Wiley & Sons, Ltd. Published 2015 by John Wiley & Sons, Ltd.

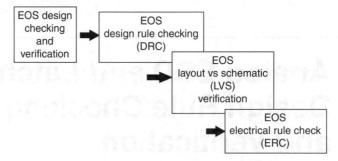

Figure 12.1 Checking and verification.

12.2.1 ESD Design Rule Checking

In semiconductor chip design, ESD design rule checking (DRC) is presently part of the ESD design discipline. In the 1980s, no ESD design rules existed in the majority of corporations. In some of the first corporations, in the 1990s, ESD design rules were first integrated into the methodology of DRC where physical dimensions could be verified.

In ESD design rule development, the rules are defined to improve and to satisfy the ESD specifications for the qualification and release of the semiconductor components (Figure 12.2). The ESD specifications of interest for the qualification and release of

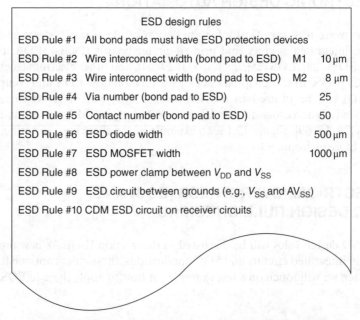

Figure 12.2 Electrostatic discharge (ESD) design rule check.

components from manufacturing are the human body model (HBM) and charged device model (CDM).

The ESD design rules are as follows [1–3]:

- Existence rule of an ESD network on all bond pads
- Wire width rule for interconnects between bond pads and ESD networks
- Number of contacts for interconnects between bond pads and ESD networks
- Number of vias for interconnects between bond pads and ESD networks
- Wire width rule for interconnects between ESD networks and power rails
- Number of contacts for interconnects between ESD networks and power rails
- Number of vias for interconnects between ESD networks and power rails
- ESD network for HBM specification compliance
- ESD network series resistor rule for receiver networks
- ESD network for CDM specification compliance
- ESD HBM network device width
- ESD CDM network device width
- ESD power clamp between power and ground (e.g., V_{DD} and V_{SS})
- ESD rail-to-rail networks between ground rails (e.g., V_{SS} to V_{SS})
- ESD network-to-ESD power clamp placement rule
- Cross-domain ESD network
- Cross-domain signal line ESD network between domains (e.g., analog-to-digital signal line)
- Differential pair pin-to-pin ESD network

12.2.2 Electrostatic Discharge Layout-versus-Schematic Verification

In ESD checking and verification, layout versus schematic (LVS) is used to insure proper implementation [1–28]. The LVS verification is a class of EDA verification software. The LVS verification is used to insure that there is proper correspondence between a design layout and a circuit schematic within an integrated circuit chip. LVS checking software recognizes the layout of a particular design and evaluates the drawn shapes of the layout; these drawn shapes represent the electrical components of the circuit and connectivity. LVS verification involves three steps: extraction, reduction, and comparison.

For ESD protection, LVS can verify the correct physical layout that is utilized for a given ESD network; this can be achieved through netlist verification. In some semiconductor chip implementation, the incorrect layout is used for the ESD network; this can lead to failure of the desired ESD specification objectives.

LVS can also be used to insure the ESD network is connected to a given signal or power pin. In many chip designs, there are cases where the design team does not place an ESD network on a given signal pin. Additionally, receivers may require two networks—one for HBM and machine model (MM) objectives and a second network for CDM protection. For sensitive receiver networks, this can lead to failing all specification, or passing one, but not the other.

It is also important to insure the correct ESD network is applied to the correct circuit. In many complex applications, mismatch between the circuit and the corresponding ESD network can occur. Using the incorrect ESD network for a corresponding signal pin can lead to functional test failure, or ESD failure.

12.2.3 ESD Electrical Rule Check (ERC)

In semiconductor chip design, ESD electrical rule check (ERC) is presently part of the ESD design discipline to insure proper operation of the ESD networks to shunt current to power or ground and to establish a low impedance alternate current path for discharge of the ESD current. Providing good ESD protection can lead to improvements to EOS by discharging EOS current through the semiconductor chip.

In ESD ERC development, the rules are focused on the resistance. By providing high series resistance, ESD current-limiting protection solutions can prevent failure of circuitry on the signal path. By providing a low resistance in the ESD network, wiring, and power rails, ESD current can be shunted through an alternate current path for discharge of the ESD current. The ESD specifications of interest for the qualification and release of components from manufacturing are HBM and CDM.

The ESD design rules are as follows:

- Wire resistance rule for interconnects between bond pads and ESD networks
- Contact resistance for interconnects between bond pads and ESD networks
- Via resistance for interconnects between bond pads and ESD networks
- Power rail wire resistance rule for interconnects between ESD networks and power rails
- Power rail contact resistance for interconnects between ESD networks and power rails
- Power rail via resistance for interconnects between ESD networks and power rails
- ESD network series resistor resistance rule for receiver networks
- ESD HBM network device resistance
- ESD CDM network device resistance
- Cross-domain signal line ESD network series resistance (e.g., analog-to-digital signal line)

12.3 ELECTRICAL OVERSTRESS (EOS) ELECTRONIC DESIGN AUTOMATION

EDA tools are very important for evaluation of EOS robustness of systems [1]. For EOS, one must evaluate the EOS protection devices, the PCB, and the ESD protection scheme. In the following sections, EOS protection networks and PCB design rules and checks are discussed.

12.3.1 Electrical Overstress (EOS) Design Rule Checking

EOS robustness can be achieved through DRC of all elements in the system [1]. This includes the EOS protection elements, the PCB, and the components themselves.

EOS DRC can be established in the EOS elements themselves. Examples of DRC rules for EOS protection devices can include (Figure 12.3):

- EOS protection device width
- EOS protection device ground rule dimension checks
- EOS protection device interconnect wiring
- EOS protection device internal bond pads

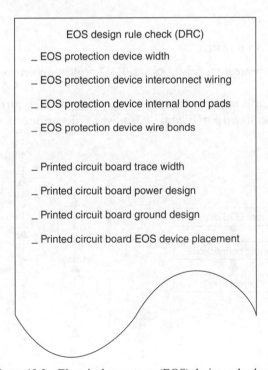

Figure 12.3 Electrical overstress (EOS) design rule check.

- EOS protection device wire bonds

- EOS protection device parasitic evaluation

EOS DRC can also be applied to the PCB design rules; this will be highlighted in a later section. EOS protection rules can be also developed for the onboard components to check to see if the components can survive the voltage and current condition of EOS events.

12.3.2 Electrical Overstress (EOS) Layout-versus-Schematic (LVS) Verification

In EOS, checking and verification is important to avoid functionality concerns, field failure, or integration errors. EOS LVS checking and verification can be used to insure proper implementation [1].

The LVS verification can insure that there is proper correspondence between a design layout and a circuit schematic within an integrated circuit chip. For EOS, LVS checking software can be used for proper integration of PCBs, EOS protection devices, and on-chip ESD protection. What is unique to EOS events, the cards, PCB, and components require verification.

EOS LVS checking and verification can address the following design errors:

- Electrical shorts

- Electrical opens

- Electrical parameter mismatch

- Mismatch of EOS protection networks and ESD protection networks

In EOS verification, it is important to check and verify that the correct EOS protection device is on the correct signal pin (Figure 12.3) as well as device directionality (Figure 12.4).

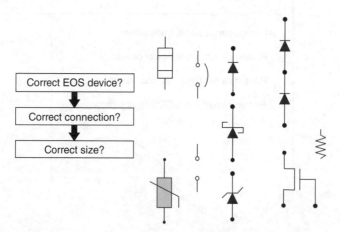

Figure 12.4 Checking and verification of correct EOS protection device.

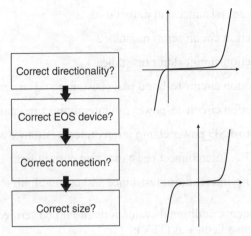

Figure 12.5   Checking and verification of correct EOS protection device directionality.

EOS protection circuits can be diodes, Schottky diodes, Zener diodes, varistors, gas discharge tubes, and a plethora of other protection schemes [1]. First, it is important to verify that the EOS protection device is a valid supported device. Second, it is important to verify that the one chosen satisfies the current-limiting resistance, or voltage trigger level. Additionally, it is important to verify that the element is suitable for the application.

Another key issue is the EOS protection device directionality. The EOS element directionality must also be suitable for the signal application. As part of the LVS checking, it is important that the correct nodes of the EOS element are the desired orientation and that electrodes are not reversed (Figure 12.5).

12.3.3 Electrical Overstress (EOS) Electrical Rule Check (ERC)

EOS ERC can provide a means of checking and verification of conformance to EOS product specification [1]. An EDA methodology can reduce the risk of field failures and field returns. For EOS events, it is necessary to evaluate the semiconductor chip, package, and PCB for checking and verification.

An EOS ERC methodology can include the following items:

- PCB trace resistance

- PCB trace maximum allowed current magnitude

- PCB components series resistance

- EOS current-limiting protection device resistance

- EOS voltage clamp protection device shunt resistance

- Component pin resistance

- Component bond wire maximum allowed current magnitude

- Component bond wire resistance and inductance
- On-chip ESD protection circuit series resistance
- On-chip ESD protection circuit shunt resistance
- On-chip ESD protection circuit-to-bond pad resistance and current limit
- On-chip ESD protection circuit-to-power rail interconnect resistance and current limit
- On-chip power rail-to-ESD power clamp interconnect resistance and current limit
- On-chip power rail V_{DD} interconnect resistance and current limit
- On-chip power rail V_{SS} interconnect resistance and current limit

For 2.5-D and 3-D systems, additional evaluations should be verified that include silicon interposers and through-silicon vias (TSV):

- Silicon interposer trace resistance and allowed current-carrying capability
- TSV resistance

12.3.4 Electrical Overstress Programmable Electrical Rule Check

EOS programmable ERC (PERC™) can provide a means of checking and verification of conformance to EOS product specification [25]. An EDA methodology can reduce the risk of field failures and field returns.

The desired lifetime of the circuit and the type of environment(s) in which the circuit will operate are important for reliability assessment and evaluation of EOS. EOS events are a function of the external temperature, electrical, magnetic, and mechanical stresses to which the circuit will be susceptible. Calibre® PERC can be used to prevent electrical circuit failure due to EOS and improve overall circuit reliability [25].

Using topological verification methods, Calibre PERC can be used to insure that these ESD structures and devices have been implemented properly. Additionally, the interconnect wire width can be verified to insure no ESD or EOS failures. For EOS events, the tool can also provide voltage propagation into the circuit to evaluate overvoltage stress from residual currents in a design [25].

12.4 PRINTED CIRCUIT BOARD (PCB) DESIGN RULE CHECKING AND VERIFICATION

In the design of the PCB, EDA design rules can be established associated with EOS, electromagnetic interference (EMI), and the EMC characteristics. EOS DRC can be easily established for the following [1]:

- Number of design levels
- Power plane design

- Ground plane design
- Signal plane design
- Trace line impedance
- Copper routing thickness
- Trace width
- Component placement
- Signal trace routing

Electronic design rule checks can be developed to address physical placement. Some design checks for placement and component selection are listed below [1]:

- Connector edge placement check
- Connector corner placement check
- Common connector
- Connector to onboard I/O components spacing check
- Connector and I/O to onboard non-I/O components spacing check
- Clock and clock oscillator placement check

Trace routing and power/ground plane decisions are key to avoid EMI, EMC, and EOS concerns.
    These can be checked and verified using EDA tools:

- Power trace EOS width
- Ground trace EOS width
- Signal trace EOS width
- Critical signal trace placement
- Non-I/O trace placement
- Non-I/O trace placement and I/O components
- Signal trace and power plane separation
- Signal trace and ground plane separation
- I/O to connector trace length
- Trace length
- Trace to board edge
- Differential signal trace pairs

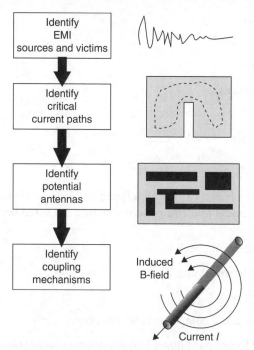

Figure 12.6   PCB design checking steps.

More sophisticated design tools can evaluate many design variables in PCB design. EDA tools can address the following (Figure 12.6):

• Identify EMI sources and victims

• Identify critical paths

• Identify potential antennae issues

• Identify coupling mechanisms

12.5 ELECTRICAL OVERSTRESS AND LATCHUP DESIGN RULE CHECKING (DRC)

EOS and latchup DRC are important to avoid electrical failure of components and PCBs. Latchup events can lead to melting of the component and its packaging. Hence, in providing an EOS robust solution, it is important to understand the limitation associated with CMOS latchup.

12.5.1 Latchup Design Rule Checking

EOS can lead to latchup in semiconductor components and systems. Latchup tolerance can be minimized through semiconductor chip process technology, design layout, circuit design, and system design. EDA can be used to check and verify latchup robustness of a component on a semiconductor chip level [11, 29–47].

Latchup DRC is important to avoid latchup concerns in semiconductor components. In the 1980s, semiconductor corporations and foundries had few means of using EDA tools to check and verify for latchup. The first latchup sections were being placed in technology design manuals in the mid-1980s [32, 33].

Today, latchup DRC are included within the design checking and verification. Latchup can occur in semiconductor devices in the following categories:

- Latchup between devices within a common circuit
- Latchup between devices between different circuits
- I/O circuit to I/O circuit latchup
- I/O circuit to ESD circuit latchup
- I/O circuit to core circuit latchup
- Domain-to-domain latchup

To address these issues, the solutions from a layout and design perspective can be achieved through physical spacing, separation of domains, and placement of guard rings to avoid formation of a parasitic pnpn that can lead to latchup. Latchup DRC categories include the following (Figure 12.7) [11, 29–54]:

- Placement of connections to power rails relative to devices
- Device-to-device placement
- Circuit-to-circuit placement
- Existence of guard ring structures and spacing around devices
- Existence of guard ring structures and spacing between devices

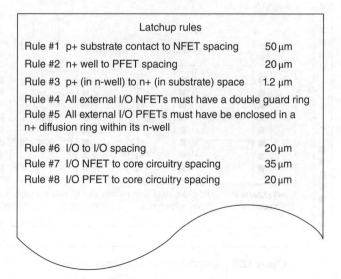

Latchup rules

Rule #1 p+ substrate contact to NFET spacing   50 µm

Rule #2 n+ well to PFET spacing   20 µm

Rule #3 p+ (in n-well) to n+ (in substrate) space   1.2 µm

Rule #4 All external I/O NFETs must have a double guard ring

Rule #5 All external I/O PFETs must have be enclosed in a n+ diffusion ring within its n-well

Rule #6 I/O to I/O spacing   20 µm

Rule #7 I/O NFET to core circuitry spacing   35 µm

Rule #8 I/O PFET to core circuitry spacing   20 µm

Figure 12.7   Latchup DRC.

• Existence of guard ring structures and spacing between circuits

• Existence of guard ring structures and spacing between separate power domains

• Existence of guard ring structures and spacing between semiconductor chip cores

In CMOS latchup, the physical dimensions associated with the parasitic pnpn network are checked and verified (Figure 12.8). CMOS latchup is a function of four fundamental variables. In a DRC, these are checked independently. In addition, there are guard ring rules. Fundamental DRC includes the following [11, 29–54]:

Minimum p+ to n-well space: The physical space between the p+ diffusion and the n-well edge is set to some minimum value based on the desired p+/n+ minimum rule.

Minimum n+ to n-well space: The physical space between the n+ diffusion and the n-well edge is set to some minimum value based on the desired p+/n+ minimum rule.

Maximum n-well resistance requirement: The maximum n-well resistance is established based on the maximum allowed well shunt resistance. This is typically represented as a physical distance between the n-well contact and the p-channel MOSFET or any p-doped element in an n-well region.

Maximum p-substrate resistance requirement: The maximum substrate resistance is established based on the maximum allowed substrate shunt resistance. This is typically represented as a physical distance between the p-well contact and the n-channel MOSFET or any n-doped element in a p-well region.

Guard ring-type rule: Design rules require guard rings for all elements electrically connected to an external node. The type of guard rings is a function of whether an element is p-doped or n-doped, and the technology requirements. Typically, n-well guard rings are placed around n-diffusions. In many technologies, double guard rings

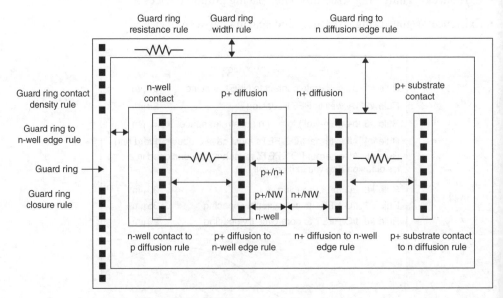

Figure 12.8 Ground rules for internal latchup.

are used (e.g., a p+ substrate ring as well as an n-well guard ring). In BiCMOS technologies, deep trench (DT) can be utilized to improve both CMOS latchup robustness and noise injection.

Minimum guard ring space rule: Typically, the guard rings are spaced relative to the physical diffusion to allow electrical biasing without interaction. In addition, the spacing is optimized as to not be too close to elements to establish interaction but at the same time at a distance too far where they do not collect minority carrier injection.

Minimum guard ring width rule: Guard ring width influences the guard ring efficiency of a guard ring structure. Hence, many technology guidelines will define the width or choose a minimum width for guard ring designs.

Maximum guard ring resistance rule: In many technologies, either the guard ring design is defined or a maximum guard ring resistance rule is established.

Butted contact rules: In many technologies, butted contacts are desired to minimize the resistance between a contact and the device, recommending that butted contacts should be utilized to minimize latchup concerns. In other technologies, this issue is avoided.

In a CMOS technology, there are many parasitic transistors inherent in the devices. In the extraction process of parasitic transistors in a physical design, there are a large number of parasitic transistors. In a multifinger MOSFET, each independent finger of the MOSFET can appear as a multifinger emitter or multifinger collector. Given there are p emitters and q collectors, there is a potential of pq-independent bipolar transistors that can be extracted. As a result of this complexity, there must be a simplification methodology to reduce the problem to a smaller set; this is done through reduction rules. T. Li highlighted three practical reduction rule cases to simplify the extraction process. In these rules, they evaluate (i) the electrical state and (ii) the bipolar current gain [12, 13].

Figure 12.9 is a drawing of the first reduction rule. A first reduction rule is the case of the "shared emitter rule." In a shared emitter rule, each independent collector has an independent voltage state and an independent bipolar current gain. The bipolar current gain can be obtained by determining the collector area and the geometric spacing relative to the emitter of interest. In this reduction process, the collectors that are farthest away and the lowest voltages are removed [12, 13].

Figure 12.10 is a representation of the second reduction rule. A second reduction rule is the case of the "shared collector rule." In a shared collector rule, each independent emitter has an independent voltage state and an independent bipolar current gain. The bipolar current gain again can be obtained by determining the emitter area and the geometric spacing relative to the collector of interest. In this reduction process, the emitters that are farthest away and the lowest voltages are removed [13].

A third rule is a "minimum bipolar current gain rule." Given a bipolar current gain was less than a given value, the bipolar parasitic transistor is not evaluated. For example, given a bipolar current gain is less than unity, it can be removed [13].

It is clear from this framework and methodology that additional rules can be defined for pnpn elements. Hence, a CMOS CAD methodology can utilize reduction rules as a means of sorting out the important parasitic bipolar transistors for CMOS latchup.

When a secondary current source external to the circuit exists, this is referred to as "external latchup." In the case of this secondary source, CMOS latchup can occur [30, 31]. The secondary source can be injection current from other circuits or other power domains.

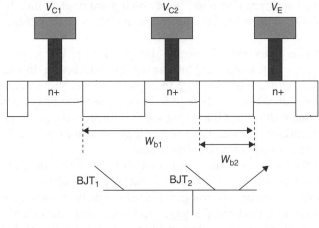

Reduction rules

If $V_{C1} < V_{C2}$ and $\beta_1 < \beta_2$ remove BJT_1

If $V_{C1} < V_{C2}$ and $\beta_1 = \beta_2$ remove BJT_1

If $V_{C1} = V_{C2}$ and $\beta_1 < \beta_2$ remove BJT_1

Figure 12.9   Latchup parasitic bipolar DRC shared emitter rule.

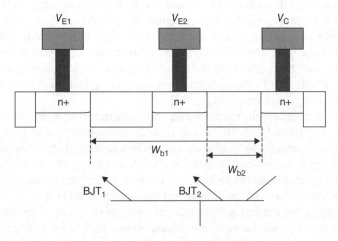

Reduction rules

If $V_{E1} = V_{E2}$ and $\beta_1 < \beta_2$ remove BJT_1

If $V_{E1} > V_{E2}$ and $\beta_1 = \beta_2$ remove BJT_1

If $V_{E1} > V_{E2}$ and $\beta_1 < \beta_2$ remove BJT_1

Figure 12.10   Latchup parasitic DRC shared collector rule.

When the secondary source is adjacent to a given sensitive circuit, the latchup robustness of the circuit can be improved by decreasing the substrate and well contact spacings to a CMOS transistor.

12.5.2 Latchup Electrical Rule Check (ERC)

In the designing of semiconductor components, latchup rules are defined for layout, circuit, and electrical characteristics. Latchup involves design layout, circuit interaction, and product current and voltage conditions [30, 31].

Latchup ERC is introduced to provide prevention of latchup due to excessive resistance [30, 31]. ERC can be introduced for a few key areas:

- N-well contact to p-channel MOSFET resistance

- P-well or p-substrate contact to n-channel MOSFET resistance

- Guard ring resistance

12.5.2.1 N-Well Contact to P-Channel MOSFET Resistance

An n-well resistor exists between the power supply (V_{DD}) and the p-channel MOSFET channel region. The n-well resistance is a function of the spacing between the physical contact to the n-well region and the p-channel MOSFET, the width of the contact region, the width of the MOSFET, and the n-well sheet resistance. For prevention of CMOS latchup, it is best to lower this value to prevent forward biasing of the p-diffusion to n-well junction.

12.5.2.2 P-Well or P-Substrate Contact to N-Channel MOSFET Resistance

A p-well resistor exists between the power supply (V_{SS}) and the n-channel MOSFET channel region. The p-well resistance is a function of the spacing between the physical contact to the p-well region and the n-channel MOSFET, the width of the contact region, the width of the MOSFET, and the p-well sheet resistance. For prevention of CMOS latchup, it is best to lower this value to prevent forward biasing of the n-diffusion to p-well junction.

12.5.2.3 Guard Ring Resistance

Guard rings are used to collect the minority carrier injection into the well and substrate regions of a semiconductor device. Guard rings constructed from metallurgical junctions allow collection of the minority carriers to prevent interaction with other devices, circuits, domains, or cores. For the guard rings to be effective, the electrical resistance between the point of collection and the power rails will be required to be low enough for efficient collection and prevention of forward biasing of the guard ring-to-substrate or guard ring-to-well metallurgical junction.

In the integration of guard rings into a semiconductor chip design, CAD methods are being used for latchup. Guard rings are placed around injection sources that can trigger latchup. Injection sources can be n+ diffusions, n-type resistors, and n-wells (e.g., in a p-type substrate wafers). One of the primary design issues with guard rings is the effectiveness to collect the minority carrier injection. The guard ring effectiveness is dependent on the guard ring type, the guard ring depth, guard ring physical width, and relative spacing from the injection source. Guard ring resistance is also a key criterion. The reasons this is a growing issue are the CMOS and BiCMOS dimensional scaling and increase in the peripheral I/O density (e.g., I/O book width scaling). With the technology dimensional scaling, the physical dimensions of the p+, n+, and n-well have been reduced. With the scaling of the minimum n-well widths, the width of n-well guard ring has been scaled. As a result, the resistance along the length has increased. With the increase in circuit density, the I/O circuit density has increased. ASIC environments have focused on reducing the width of the I/O peripheral book to allow more I/O circuits on the periphery of a semiconductor chip. In this process, the peripheral I/O length has been increased to compensate the reduction of the peripheral I/O book width.

With the placement of an n-type guard ring in the p-type substrate, a metallurgical junction is formed which can collect the minority carrier electrons injected into the substrate. As an example, an n-type guard ring is biased to the power supply voltage, collecting the injection current.

At low injection currents, the electrons are collected by the reverse-biased metallurgical junction formed between the substrate and the n-type guard ring. But, at very high injection currents, the series resistance between the power supply voltage and the guard ring is a key latchup design factor.

The injection source serves as an "emitter," the substrate serves as a "base" region, and the guard ring serves as a "collector." When the emitter–base junction is forward active, the electrons are injected into the substrate region. When the collector is biased positive at the power supply voltage, the collector-to-emitter voltage is positive. In this state, the parasitic transistor formed between the injection source and the guard ring is forward active. When the resistance of the guard ring increases, a voltage drop occurs in the guard ring. The voltage drop is equal to the product of the guard ring resistance and the injection current:

$$\Delta V = I_{inj} R_{GR}$$

where I_{inj} is the injection current and R_{GR} is the guard ring resistance between the point of injection and the power supply voltage. At the location of the injection, the voltage at the guard ring is equal to

$$V_{GR} = V_{DD} - \Delta V_{GR} = V_{DD} - I_{inj} R_{GR}$$

As the voltage drop increases due to the injection current, the guard ring voltage at the point of injection will decrease. When the effectiveness of the guard ring to collect the current is minimized as a result of the debiasing, the minority carrier electron current will flow to alternative structures (e.g., outside of the guard ring). Design parameters that influence the resistance are the following:

- Guard ring sheet resistance (e.g., n-well sheet resistance or plurality of implants in the guard ring)

- Guard ring width

- Guard ring contact density

- Guard ring contact resistance

- Guard ring silicide resistance

- Metal bus resistance

- Distance between the injection location and the power supply voltage source

Historically, guard ring resistance was not a critical issue due to the technology, the ground rule dimensions, and the I/O density. From the 1980s to 2000, the guard rings used were typically n-well regions. In this time frame, the ground rules for both diffused and retrograde wells prevented narrow width n-well regions. As a result, the ground rules prevented scaling of the guard ring widths below some minimum dimension (e.g., typically, wells could not be scaled below 3–7 μm). With the utilization of the n-diffusion, silicides (e.g., titanium silicide and cobalt silicide), and large contact dimensions, the resistance was very low. Additionally, due to wide "wiring tracks" and peripheral I/O design, the power bus width was wide (e.g., 10–30 μm). In addition, the I/O density was low.

In this millennium, the vertical semiconductor process profile was scaled, leading to higher well sheet resistance. In addition, vertical scaling allowed for a decreased minimum well width requirement allowing a narrower guard ring structure in I/O design. In each technology generation, the number of I/O increases leading to high aspect ratio I/O books that are long and narrow. In this case, the guard ring width is reduced, and the length between injection sources and the power supply voltage are increased. In addition, with the metal scaling, the wire widths are reduced to allow a higher density of wire tracks. With all the scaling issues for both the semiconductor process and the semiconductor chip layout design, the resistance issue is more critical.

A CMOS latchup CAD system can be developed that addresses the guard ring resistance [30]. The CMOS latchup CAD evaluation must address a maximum resistance requirement for the guard ring resistance (Figure 12.11). The guard ring resistance can be evaluated as follows:

- Identify injection source.

- Identify the location near the guard ring structure.

- Calculate the total resistance to the power supply V_{DD}.

- Evaluate the maximum resistance allowed for the guard ring for the given conditions.

A key latchup design practice is as follows:

- Establish a guard ring structure under the condition of maximum resistance criteria.

- Establish a resistance criteria associated with a resistance calculation from the location of injection sources and its intersection at the guard ring to the power supply voltage location which "sinks" the latchup current. The effective resistance of all structures is used in the resistance calculation (e.g., sheet resistance, silicides, contact, and metal bussing).

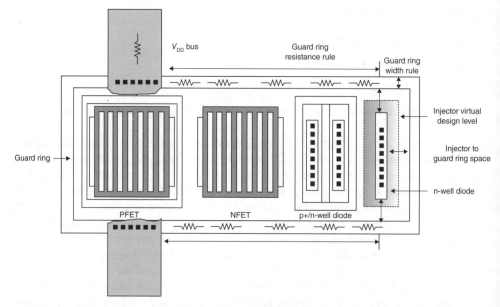

Figure 12.11 Guard ring resistance ground rule with injection source.

12.6 WHOLE-CHIP CHECKING AND VERIFICATION METHODS

Whole-chip ESD checking and verification methods have been a significant challenge in the semiconductor industry. Whole-chip ESD checking and verification has been the "holy grail" of semiconductor chip design. Today, there is still no standard checking and verification tool that is used in the foundry business [23, 55–69].

Due to Moore's law technology scaling, the number of circuits continues to increase each technology generation. With the increase in the number of circuits on a chip, the number of peripheral I/O continues to increase according to Rent's rule.

The complexity of semiconductor chips continues to increase with mixed-voltage interface (MVI), MS, SOC, and 3-D multichip environments. Technologies support digital, analog, power, and RF domains in a common chip or multichip system. With the increased complexity, multiple power and grounds exist for each domain.

The whole-chip verification methods focused on HBM and CDM ESD events. HBM methods address ESD critical discharge paths between the bond pad and power rails. These HBM-based checking and verification methods evaluate the voltage drop along the discharge path, between two pads. CDM checking and verification methods evaluate voltage clamping on each pad.

With the increase in chip size, and the number of bond pads, the run time increases. Checking and verification systems require methods to reduce the netlist and run time [64–69]. A methodology to reduce the run time was the representation of each domain as a "macro model" for each power domain was developed. A CDM event was performed on the network to focus on the cross-domain signals. A second methodology simulated CDM events on a reduced netlist and applied a random walk [61]. Another method reduces the netlist to only include elements in the ESD current path and initiate a SPICE circuit pre- and postprocess [64]. Other methodologies check all circuit elements electrically connected to bond pads. In these methods, cross-domain signal lines and interconnections of elements in these domains are not all addressed.

12.7 CROSS-DOMAIN SIGNAL LINE CHECKING AND VERIFICATION

With the increase in complexity, the number of cross-domain signal lines increases. Cross-domain signal lines are a source of ESD failures. Hence, methods are required for identifying these cross-domain signal lines where ESD failures exist internal to the semiconductor chip. New methodologies focus on these cross-domain signal lines and the internal ESD failures.

12.7.1 Cross-Domain Signal Line Checking and Verification Flow System

With the increase in complexity, the number of cross-domain signal lines increases. Cross-domain signal lines are a source of ESD failures. A full-chip ESD verification methodology that focuses on internal interfaces between power and ground is discussed by Z. Lu and D.A. Bell [62].

In this methodology, the focus uses the device netlist topology to check all domains crossing interfaces, as well as the ESD networks in the domains. The checking and verification method checks the design hierarchically. This method utilizes a "bottom-up" approach where the checking is done cell by cell; this is achieved using a novel concept of a topology-aware net type.

Figure 12.12 shows the verification flow diagram. The verification uses a SPICE netlist which is either a circuit schematic or layout based. In the chip layout netlist, this methodology must identify ESD protection networks. In this method, an ESD rule deck supports the ESD verification engine. The ESD rule deck identifies the power and ground rails, as well as ESD-related information. The hierarchical algorithm has two fundamental steps: an initialization step and rule checking step. The initialization step collects the ESD-related topology information. In the second step, the rule checking step, the algorithm checks the ESD rules, cell by cell.

In this method, there is a topology-aware net type and a topology-aware path type. The topology-aware net type recognizes that a device is contained in a cell; this allows for finding elements associated with ESD protection. Secondly, a topology-aware path type identifies the electrical current paths. These net types and net path types are propagated across the design. Figure 12.13 shows an example where the cross-domain signal has both a net type and path type associated with the signal line in both the domains.

This methodology introduces a hierarchical methodology that does not require netlist reduction. With the utilization of hierarchical method, each cell is evaluated independently.

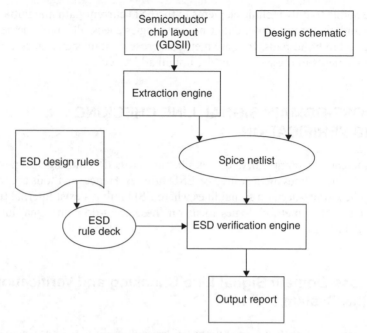

Figure 12.12 Verification flow.

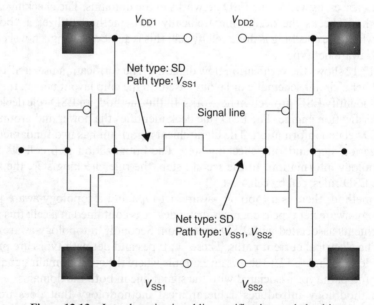

Figure 12.13 Interdomain signal line verification and checking.

12.7.2 Cross-Domain Analog Signal Line Checking and Verification Flow System

In analog design, with the complexity of the ESD library, additional requirements are needed to provide ESD checking and verification. Figure 12.14 shows an example of a different analog overall verification flow [63]. The key features of this analog ESD checker are as follows:

- Identify the ESD cell
- Verify ESD product requirement fulfillment (e.g., capability)
- Verify ESD that does not impact functionality (e.g., transparency)
- Verify existence of ESD cell (e.g., availability)

An analog design verification system must have the following characteristics:

- Support a large component portfolio
- Support a diverse component type (e.g., CMOS, LDMOS, BCD)
- Support a large ESD library

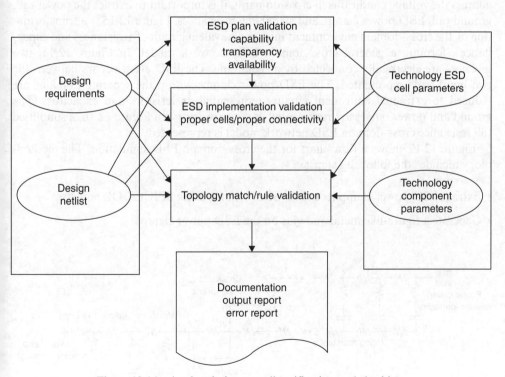

Figure 12.14 Analog design overall verification and checking.

- Check complex circuits

- Validation of layout check

The validation of the ESD cell is key to this methodology. The methodology addresses validating capability, transparency, and availability. The tool has the ability to reduce the size of the network analyzed by topology matching and introduces model simplification by the use of a "black box" path representation of ESD cells and technology components.

In this methodology, the cross-domain issues can be evaluated. This is achieved by extraction of the ESD cells and resistances. The tool identifies "at-risk constructions" where unintentional cross-domain connections exist that could lead to ESD failures. The methodology has the ability to "jump" over predefined series components to identify the "at-risk" topology across two domains.

12.7.3 Cross-Domain Checking and Verification: Resistance Extraction Methodology

Cross-domain checking and verification are important in MS semiconductor applications and can be integrated into whole-chip ESD simulation [64–69]. In this whole-chip methodology, the ESD paths are evaluated including power and ground connections. To address the voltage conditions in cross-domains, it is important to extract the power rail, ground rail, ESD power clamp, and signal line resistances. Figure 12.15 is a transformation of the cross-domain environment into an equivalent circuit of resistances and capacitances, forming a general cross-domain ESD network model. In Figure 12.15, two inverters are electrically connected by a signal line. The ESD event occurs between the two bond pads represented. The ESD power clamps within the separate domains are reduced to extracted equivalent circuits represented as resistors and capacitors. The ground and power busses are represented as series resistors. In Figure 12.16, a simplified full resistance cross-domain ESD network model is represented.

Figure 12.17 shows a flow chart for the cross-domain ESD simulation. The methodology includes the following steps:

- Extract ESD power clamp from I/O GDS without using SPICE/CDL netlists.

- Generate a figure-like metal and vias on the ESD power clamp.

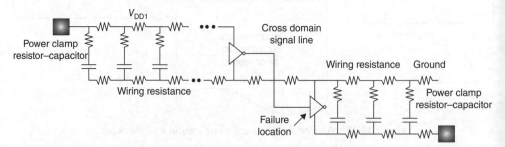

Figure 12.15   Cross-domain analysis including both equivalent resistance and capacitance.

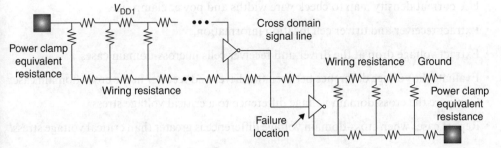

Figure 12.16   Cross-domain resistance analysis.

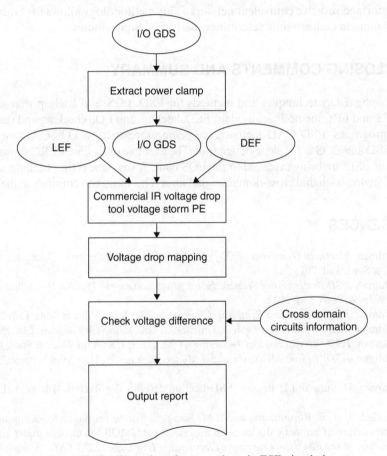

Figure 12.17   Flow chart for cross-domain ESD simulation.

- Set the resistance parameter based on the on-resistance of the ESD power clamp.

- Determine voltage drop using commercial voltage drop tool using the transformed I/O layout, LEF, and DEF.

- Generate a voltage drop map.

- Use current density map to check wire widths and power clamp size.

- Extract receiver and driver cell location information.

- Extract voltage drop at the driver and receiver cells in cross-domain cases.

- Evaluate the voltage difference between the driver and receiver in the cross-domain case.

- Compare the cross-domain voltage difference to a critical voltage stress.

- Report cases when cross-domain voltage difference is greater than critical voltage stress.

This efficient whole-chip methodology considers full ESD paths and the power and ground network through a layout-based extraction of ESD elements and converts them to a transformed resistive equivalent network. This methodology allows for evaluation of the cross-domain concern and determines the ESD design window.

12.8 CLOSING COMMENTS AND SUMMARY

In this chapter, EDA techniques and methods for ESD, EOS, and latchup were discussed. DRC, LVS, and ERC methods are used for ESD, latchup, and EOS checking and verification. As time progresses, ESD CAD methods are being propagated to EOS CAD methods, to address ESD and EOS in the same design tool. The example of Calibre PERC shows how the methods of ESD are being extended to the EOS issue. A key issue is the checking and verification of analog-to-digital cross-domain signal lines. This trend will continue in the future.

REFERENCES

1. S. Voldman. *Electrical Overstress (EOS): Devices, Circuits and Systems*. Chichester, UK: John Wiley & Sons, Ltd, 2013.
2. S. Voldman. *ESD Basics: From Semiconductor Manufacturing to Product Use*. Chichester, UK: John Wiley & Sons, Ltd, 2012.
3. S. Voldman. *ESD: Design and Synthesis*. Chichester, UK: John Wiley & Sons, Ltd, 2011.
4. S. Voldman. *ESD: Circuits and Devices*. Chichester, UK: John Wiley & Sons, Ltd, 2005.
5. S. Voldman. *ESD: RF Circuits and Technology*. Chichester, UK: John Wiley & Sons, Ltd, 2006.
6. S. Voldman. *ESD: Failure Mechanisms and Models*. Chichester, UK: John Wiley & Sons, Ltd, 2009.
7. A. Kapoor, D. Suh, and P. Bendix. ESD/latchup design rules. Technical Report, LSI Logic, 1996.
8. T. Li, C.H. Tsai, E. Rosenbaum, and S.M. Kang. Substrate resistance modeling and circuit level simulation of parasitic device coupling effects for CMOS I/O circuits under ESD stress. *Proceedings of the Electrical Overstress/Electrostatic Discharge (EOS/ESD) Symposium*, 1998; 281–289.
9. T. Li, C.H. Tsai, Y. Huh, E. Rosenbaum, and S.M. Kang. A new algorithm for circuit level electro-thermal simulation under EOS/ESD stress. *Proceedings of the IEEE International Reliability Workshop (IRW)*, 1997; 130–131.
10. T. Li, S. Ramaswamy, E. Rosenbaum, and S.M. Kang. Simulation and optimization of deep submicron output protection device. *Proceedings of the Custom Integrated Circuits Conference (CICC)*, 1997; 159–162.

11. T. Li and S.M. Kang. Layout extraction and verification methodology for CMOS I/O circuits. *Proceedings of the IEEE/ACM Design Automation Conference (DAC)*, 1998; 291–296.

12. T. Li, C.H. Tsai, E. Rosenbaum, and S.M. Kang. Substrate modeling and lumped substrate resistance extraction for latchup/ESD circuit simulations. *Proceedings of the IEEE/ACM Design Automation Conference (DAC)*, 1999; 549–554.

13. T. Li. *Design Automation for Reliable CMOS Chip I/O Circuits*. Ph.D. Thesis, University of Illinois Urbana-Champaign, UILU-ENG-98-2219, August 1998.

14. R.S. Bass, Jr., D.J. Nickel, D.C. Sullivan, and S.H. Voldman. Method of automated ESD protection level verification. U.S. Patent No. 6,086,627, July 11, 2000.

15. R.Y. Zhan, H.G. Feng, Q. Wu, G. Chen, X.K. Guan, and A.Z. Wang. A technology-independent CAD tool for ESD protection device extraction: ESD extractor. *Proceedings of the International Conference on Computer-aided Design*, 2002; 510–513.

16. P. Venugopal, S. Sinha, S. Ramaswamy, C. Duvvury, G.C. Prasad, C.S. Raghu, and G. Kadamati. Integrated circuit design error detector for electrostatic discharge and latch-up applications. U.S. Patent No. 6,493,850, December 10, 2002.

17. S. Ramaswamy, S. Sinha, G. Kadamati, and R. Gharpurey. Semiconductor device extractor for electrostatic discharge and latch-up applications. U.S. Patent No. 6,553,542, April 22, 2003.

18. R.Y. Zhan. *ESDCat: A New CAD Software Package for Full-chip ESD Protection Circuit Verification*. Doctoral Dissertation, Illinois Institute of Technology, Chicago, IL, October 2005.

19. S. Voldman, S. Strang, and D. Jordan. A design system for auto-generation of ESD circuits. *Proceedings of the International Cadence Users Group (ICUG)*, September 2002.

20. S. Voldman, S. Strang, and D. Jordan. An automated electrostatic discharge computer-aided design (CAD) system with the incorporation of hierarchical parameterized cells in BiCMOS analog and RF technology for mixed signal applications. *Proceedings of the Electrical Overstress/Electrostatic Discharge (EOS/ESD) Symposium*, October 2002; 296–305.

21. S. Voldman. Automated hierarchical parameterized ESD network design and checking system. U.S. Patent No. 6,704,179, March 9, 2004.

22. D.S. Collins, D.L. Jordan, S.E. Strang, and S. Voldman. ESD design, verification, and checking system and method of use. U.S. Patent No. 7,134,099, November 7, 2006.

23. T. Smedes, N. Trivedi, J. Fleurimont, A.J. Huitsing, P.C. de Jong, W. Scheucher, and J. Van Zwol. A DRC-based check tool for ESD layout verification. *Proceedings of the Electrical Overstress/Electrostatic Discharge (EOS/ESD) Symposium*, 2009; 292–300.

24. EDA Tool Working Group. *ESD Electronic Design Automation Checks*. ESD TR-18.0-01-11, http://www.esda.org/Documents.html.

25. C. Robertson. Calibre PERC: Preventing electrical overstress failures. *EE Times Design, EE Times*, February 9, 2012.

26. TSMC Open Innovation Platform Ecosystem Forum. Improving analog/mixed signal circuit reliability at advanced nodes. *EDA Technical Presentations*, 2011.

27. M. Khazhinsky. ESD electronic design automation checks. *In Compliance Magazine*, August 2012.

28. M. Hogan. Electronic design automation checks, Part II: Implementing ESD EDA checks in commercial tools. *In Compliance Magazine*, 2012.

29. R. Troutman. *CMOS Latchup in Semiconductor Technology: The Problem and the Cure*. New York: Kluwer Academic Publications, 1985.

30. S. Voldman. *Latchup*. Chichester, UK: John Wiley & Sons, Ltd, 2007.

31. M.D. Ker and S.F. Hsu. *Transient-Induced Latchup in CMOS Integrated Circuits*. Singapore: John Wiley & Sons (Asia) Pte, Ltd, 2009.

32. S. Voldman. *Latchup Design Rules Section*, CMOS IV Technology Manual. Burlington, VT: IBM Corporation, 1984.

33. S. Voldman. *Latchup Design Rules Section*, CMOS V Technology Manual. Burlington, VT: IBM Corporation, 1988.

34. T. Li, Y. Huh, and S.M. Kang. Automated extraction of parasitic BJTs for CMOS I/O circuits under ESD stress. *Proceedings of the IEEE International Integrated Reliability Workshop (IRW)*, 1997; 103–109.

35. S. Voldman, C.N. Perez, and A. Watson. Guard rings: Structures, design methodology, integration, experimental results and analysis for RF CMOS and RF mixed signal BiCMOS silicon germanium technology. *Journal of Electrostatics*, **64**, 2006; 730–743.

36. C.N. Perez and S. Voldman. Method of forming guard ring parameterized cell structure in a hierarchical parameterized cell design, checking and verification system. U.S. Patent Application No. 20,040,268,284, December 30, 2004.

37. M.S. Galland, P.A. Habitz, and S.E. Washburn. Method and apparatus for detecting devices that can latchup. U.S. Patent No. 6,848,089, January 25, 2005.

38. M.D. Ker, H.C. Jiang, J.J. Peng, and T.L. Shieh. Automatic methodology for placing the guard rings into a chip layout to prevent latchup in CMOS IC's. *Proceedings of the 8th IEEE International Conference on Electronics, Circuits and Systems (ICECS)*, 2001; 113–116.

39. S. Kimura and H. Tsujikawa. Latch-up verifying method and latch-up verifying apparatus capable of varying over-sized region. U.S. Patent No. 6,490,709, December 3, 2002.

40. S. Kimura and H. Tsujikawa. Latch-up verifying method and latch-up verifying apparatus capable of varying over-sized region. U.S. Patent No. 6,718,528, April 6, 2004.

41. S. Voldman. Latch-up analysis and parameter modification. U.S. Patent No. 6,996,786, February 7, 2006.

42. A.E. Watson and S. Voldman. Method of quantification of transmission probability for minority carrier collection in a semiconductor chip. U.S. Patent No. 7,200,825, April 3, 2007.

43. S. Voldman. Methodology for placement based on circuit function and latchup sensitivity. U.S. Patent No. 7,089,520, August 8, 2006.

44. J.M. Cohn. *Automatic Device Placement for Analog Cells in KOAN*. Ph.D. Thesis, Department of Electrical Engineering and Computer Science, Carnegie Mellon University, Pittsburgh, PA, February 1992.

45. J.M. Cohn, D.J. Garrod, R.A. Rutenbar, and L.R. Carley. KOAN/ANAGRAM II: New tools for device-level analog placement and routing. *IEEE Journal of Solid State Circuits*, **SS-26** (3), March 1991; 330–342.

46. B. Basaran. *Latchup-Aware Placement and Parasitic-Bounded Routing of Custom Analog Cells*. M.S. Thesis, Department of Electrical Engineering and Computer Science, Carnegie Mellon University, Pittsburgh, PA, May 7, 1993.

47. B. Basaran, R. Rutenbar, and L. Carley. Latchup-aware placement and parasitic-bounded routing of custom analog cells. *Proceedings of the 1993 IEEE/ACM International Conference on Computer Aided Design (ICCAD)*, 1993; 415–421.

48. S. Voldman. Structure, structure and method of latch-up immunity for high and low voltage integrated circuits. U.S. Patent No. 8,519,402, August 27, 2013.

49. S. Voldman. Semiconductor structure and method of designing semiconductor structure to avoid high voltage initiated latch-up in low voltage sectors. U.S. Patent No. 8,423,936, August 27, 2013.

50. S. Voldman. Guard ring structures for high voltage CMOS/low voltage CMOS technology using LDMOS (lateral double-diffused metal oxide semiconductor) device fabrication. U.S. Patent No. 8,110,853, February 7, 2012.

51. S. Voldman. Structure and method for latchup improvement using wafer via latchup guard ring. U.S. Patent No. 7,989,282, August 2, 2011.

52. S. Voldman. Structure and method for latchup improvement using wafer via latchup guard ring. U.S. Patent No. 8,390,074, March 5, 2013.

53. P. Chapman, D.S. Collins, and S. Voldman. Structure and method for latchup robustness with placement of through wafer via within CMOS circuitry. U.S. Patent No. 8,420,518, April 16, 2013.

54. P. Chapman, D.S. Collins, and S. Voldman. Structure and method for latchup robustness with placement of through wafer via within CMOS circuitry. U.S. Patent No. 8,017,471, September 13, 2011.

55. M. Ker, C. Wu, H. Chang, and T. Wu. Whole chip ESD protection scheme for CMOS mixed-mode IC's in deep-submicron CMOS technology. *Proceedings of the IEEE Custom Integrated Circuit Conference (CICC)*, 1997; 31–34.

56. S. Sinha, H. Swaminathan, G. Kadamati, and C. Duvvury. An automated tool for detecting ESD design errors. *Proceedings of the Electrical Overstress/Electrostatic Discharge (EOS/ESD) Symposium*, 1998; 208–217.

57. M. Baird and R. Ida. VerifyESD: A tool for efficient circuit level ESD simulations of mixed signal ICs. *Proceedings of the Electrical Overstress/Electrostatic Discharge (EOS/ESD) Symposium*, 2000; 465–469.

58. P. Ngan, R. Gramacy, C.K. Wong, D. Oliver, and T. Smedes. Automatic layout based verification of electrostatic discharge paths. *Proceedings of the Electrical Overstress/Electrostatic Discharge (EOS/ESD) Symposium*, 2001; 96–101.

59. R. Zhan, H. Xie, H. Feng, and A. Wang. ESDZapper: A new layout-level verification tool for finding critical discharging path under ESD stress. *Proceedings of the ASPDAC*, 2005; 79–82.

60. H.Y. Liu, C.W. Lin, S.J. Chou, W.T. Tu, C.H. Liu, Y.W. Chang, and S.Y. Kuo. Current path analysis for electrostatic discharge protection. *Proceedings of the ICCAD*, 2006; 510–515.

61. J. Connor, S. Mitra, G. Wiedemeir, A. Wagstaff, R. Gauthier, M. Muhammad, and J. Never. ESD simulation using fully extracted netlist to validate ESD design improvement. *Proceedings of the International ESD Workshop (IEW)*, 2007; 396–407.

62. Z.Y. Lu and D.A. Bell. Hierarchical verification of chip-level ESD design rules. *Proceedings of the Electrical Overstress/Electrostatic Discharge (EOS/ESD) Symposium*, 2010; 97–102.

63. H. Kunz, G. Boselli, J. Brodsky, M. Hambardzumyan, R. Eatmon. An automated ESD verification tool for analog design. *Proceedings of the Electrical Overstress/Electrostatic Discharge (EOS/ESD) Symposium*, 2010; 103–110.

64. M. Okushima, T. Kitayama, S. Kobayashi, T. Kato, and M. Hirata. Cross domain protection analysis and verification using whole chip ESD simulation. *Proceedings of the Electrical Overstress/Electrostatic Discharge (EOS/ESD) Symposium*, 2010; 119–125.

65. M. Muhammad, R. Gauthier, J. Li, A. Ginawi, J. Montstream, S. Mitra, K. Chatty, A. Joshi, K. Henderson, N. Palmer, and B. Hulse. An ESD design automation framework and tool flow for nano-scale CMOS technology. *Proceedings of the Electrical Overstress/Electrostatic Discharge (EOS/ESD) Symposium*, 2010; 91–96.

66. G.C. Tian, Y.P. Xiao, D. Connerney, T.H. Kang, A. Young, and Q. Liu. A predictive full chip dynamic ESD simulation and analysis tool for analog and mixed signal ICs. *Proceedings of the Electrical Overstress/Electrostatic Discharge (EOS/ESD) Symposium*, 2011; 285–293.

67. N. Chang, Y.L. Liao, Y.S. Li, P. Johari, and A. Sarkar. Efficient multi-domain ESD analysis and verification for large SOC designs. *Proceedings of the Electrical Overstress/Electrostatic Discharge (EOS/ESD) Symposium*, 2011; 300–306.

68. N. Trivedi, H. Gossner, H. Dhakad, B. Stein, and J. Schneider. An automated approach for verification of on-chip interconnect resistance for electrostatic discharge paths. *Proceedings of the Electrical Overstress/Electrostatic Discharge (EOS/ESD) Symposium*, 2011; 307–314.

69. H. Marquardt, H. Wagieh, E. Weidner, K. Domanski, and A. Ille. Topology-aware ESD checking: A new approach to ESD protection. *Proceedings of the Electrical Overstress/Electrostatic Discharge (EOS/ESD) Symposium*, 2012; 85–90.

Appendix: Standards

ESD ASSOCIATION

ANSI/ESD S1.1-2006. Wrist Straps.

ESD DSTM 2.1. Garments.

ANSI/ESD STM 3.1-2006. Ionization.

ANSI/ESD SP3.3-2006. Periodic Verification of Air Ionizers.

ANSI/ESD STM 4.1-2006. Worksurfaces—Resistance Measurements.

ANSI/ESD STM 3.1-2006. ESD Protective Worksurfaces—Charge Dissipation Characteristics.

ANSI/ESD STM 5.1-2007. Electrostatic Discharge Sensitivity Testing—Human Body Model (HBM) Component Level.

ANSI/ESD STM 5.1.1-2006. Human Body Model (HBM) and Machine Model (MM) Alternative Test Method: Supply Pin Ganging—Component Level.

ANSI/ESD STM 5.1.2-2006. Human Body Model (HBM) and Machine Model (MM) Alternative Test Method: Split Signal Pin—Component Level.

ANSI/ESD S5.2-2006. Electrostatic Discharge Sensitivity Testing—Machine Model (MM) Component Level.

ANSI/ESD S5.3.1-2009. Charged Device Model (CDM)—Component Level.

ANSI/ESD SP5.3.2-2008. Electrostatic Discharge Sensitivity Testing—Socketed Device Model (SDM) Component Level.

ANSI/ESD STM 5.5.1-2008. Electrostatic Discharge Sensitivity Testing—Transmission Line Pulse (TLP) Component Level.

ESD: Analog Circuits and Design, First Edition. Steven H. Voldman.
© 2015 John Wiley & Sons, Ltd. Published 2015 by John Wiley & Sons, Ltd.

ANSI/ESD SP5.5.2-2007. Electrostatic Discharge Sensitivity Testing—Very Fast Transmission Line Pulse (VF-TLP) Component Level.

ANSI/ESD SP6.1-2009. Grounding.

ANSI/ESD S7.1-2005. Resistive Characterization of Materials—Floor Materials.

ANSI/ESD S8.1-2007. Symbols—ESD Awareness.

ANSI/ESD STM 9.1-2006. Footwear—Resistive Characterization.

ESD SP9.2-2003. Footwear—Foot Grounders Resistive Characterization.

ANSI/ESD SP10.1-2007. Automatic Handling Equipment (AHE).

ANSI/ESD STM 11.11-2006. Surface Resistance Measurement of Static Dissipative Planar Materials.

ESD DSTM 11.13-2009. Two Point Resistance Measurement.

ANSI/ESD STM 11.31-2006. Bags.

ANSI/ESD STM 12.1-2006. Seating—Resistive Measurements.

ESD STM 13.1-2000. Electrical Soldering/Desoldering Hand Tools.

ANSI/ESD SP14.1. System Level Electrostatic Discharge (ESD) Simulator Verification.

ESD SP14.3-2009. System Level Electrostatic Discharge (ESD) Measurement of Cable Discharge Current.

ANSI/ESD SP15.1-2005. In Use Resistance Testing of Gloves and Finger Cots.

ANSI/ESD S20.20-2007. Protection of Electrical and Electronic Parts, Assemblies, and Equipment.

ANSI/ESD STM 97.1-2006. Floor Materials and Footwear—Resistance Measurements in Combination with A Person.

ESD Association. DSP 14.1-2003. *ESD Association Standard Practice for the Protection of Electrostatic Discharge Sensitive Items—System Level Electrostatic Discharge Simulator Verification Standard Practice*. Standard Practice (SP) document, 2003.

ESD Association. DSP 14.3-2006. *ESD Association Standard Practice for the Protection of Electrostatic Discharge Sensitive Items—System Level Cable Discharge Measurements Standard Practice*. Standard Practice (SP) document, 2006.

ESD Association. DSP 14.4-2007. *ESD Association Standard Practice for the Protection of Electrostatic Discharge Sensitive Items—System Level Cable Discharge Test Standard Practice*. Standard Practice (SP) document, 2007.

JEDEC

EIA JESD78. *IC Latchup Test*, JEDEC Organization, 2010.

INTERNATIONAL ELECTRO-TECHNICAL COMMISSION (IEC)

IEC 61000-4-2. Electromagnetic compatibility (EMC): Testing and measurement techniques—Electrostatic discharge immunity test, *IEC International Standard*, 2001.

IEC 61000-4-2. Electromagnetic compatibility (EMC)—Part 4-2: Testing and measurement techniques—Electrostatic discharge immunity test, *IEC International Standard*, 2008.

IEC 61000-4-5. Electromagnetic compatibility (EMC)—Part 4-5: Testing and measurement techniques—Surge immunity test, *IEC International Standard*, 2000.

IEEE

IEEE Standard C62.45-1991. IEEE Guide on Surge Testing for Equipment Connected to Low-Voltage AC Power Circuit, 1992.

DEPARTMENT OF DEFENSE (DOD)

DOD-HDBK-263. Electrostatic Discharge Control Handbook for Protection of Electrical and Electronic Parts, Assemblies and Equipment.

DOD-STD-1686. Electrostatic Discharge Control Program for Protection of Electrical and Electronic Parts, Assemblies and Equipment.

DOD-STD-2000-2A. Part and Component Mounting for High Quality/High Reliability Soldered Electrical and Electronic Assembly.

MILITARY STANDARDS

MIL-STD-454. Standard General Requirements for Electronic Equipment.

MIL-STD-461E. Requirements for the Control of Electromagnetic Interference Characteristics of Subsystems and Equipment, August 20, 1999.

MIL-STD-785. Reliability Program for System and Equipment Development and Production.

MIL-STD-883. Method 3015-4—Electrostatic Discharge Sensitivity Classification.

MIL-STD-1686A. Electrostatic Discharge Control Program for Protection of Electrical and Electronic Parts, Assemblies and Equipment.

MIL-E-17555. Electronic and Electrical Equipment, Accessories, and Provisioned Items (Repair Parts: Packaging of).

MIL-D-81997. Pouches, Cushioned, Flexible, Electrostatic Free, Reclosable, Transparent.

MIL-D-82646. Plastic Film, Conductive, Heat Sealable, Flexible.

MIL-D-82647. Bags, Pouches, Conductive, Plastic, Heat Sealable, Flexible.

IEC 801-2. Electromagnetic Compatibility for Industrial Process Measurements and Control Equipment, Part 2: Electrostatic Discharge (ESD) Requirements.

EIA-541. Packaging Material Standards for ESD Sensitive Materials.

JEDEC 108. Distributor Requirements for Handling Electrostatic Discharge Sensitive (ESDS) Devices.

SAE

SAE J551. Performance Levels and Methods of Measurement of Electromagnetic Compatibility of Vehicles and Devices (60 Hz to 18 GHz), Society of Automotive Engineers, June 1996.

SAE J1113. Electromagnetic Compatibility Measurement Procedure for Vehicle Component (Except Aircraft) (60 Hz to 18 GHz), Society of Automotive Engineers, June 1995.

Appendix: Glossary of Terms

Analog-to-Digital (ADC) Converter   a voltage converter that whose input signals are analog signals and whose output signal is converted to a digital signal.

Audits   Business processes review to verify conformance and compliance to ESD procedures and standards.

Bandgap Reference Circuit   A temperature-independent voltage reference circuit widely used in integrated circuits. It produces a fixed (constant) voltage irrespective of power supply variations, temperature changes, and the loading on the device.

Boost Converter   A DC-to-DC power converter with an output voltage greater than its input voltage. It is a class of switched-mode power supply (SMPS).

Buck Converter   A step-down DC-to-DC converter. Its design is a switched-mode power supply that uses two switches (a transistor and a diode), an inductor and a capacitor.

Buck/Boost Converter   A type of DC-to-DC converter that has an output voltage magnitude that is either greater than or less than the input voltage magnitude.

Cable Discharge Event (CDE)   An electrostatic discharge event from a cable source.

Cassette Model   A test method whose source is a capacitor network with a 10 pF capacitor. This is also known as the Small Charge Model (SCM) and the "Nintendo model."

Charged Board Event (CBE)   A test method for evaluation of the charging of a packaged semiconductor chip mounted on a board, followed by a grounding process. The semiconductor chip is mounted on a board during this test procedure. The board is placed on an insulator during this test.

Charged Device Model (CDM)   A test method for evaluation of the charging of a packaged semiconductor chip, followed by a grounding a pin. The semiconductor chip is not socketed but placed on an insulator during the test.

Circuit Breaker   An electrical over-current (EOC) protection circuit element.

Common Centroid Design   A design methodology based on symmetry of a common centroid.

Comparator Circuit   A circuit that is a device that compares two voltages or currents and outputs a digital signal indicating which is larger.

ESD: Analog Circuits and Design, First Edition. Steven H. Voldman.
© 2015 John Wiley & Sons, Ltd. Published 2015 by John Wiley & Sons, Ltd.

Conductor A material that allows free flow of electrons. Examples of conductors include metal materials such as copper and aluminum. A material whose conductivity that exceeds insulators and semiconductors.

Cuk Converter A type of DC–DC converter that has an output voltage magnitude that is either greater than or less than the input voltage magnitude.

Current Mirror A circuit that copies a current through one active device by controlling the current in another active device of a circuit providing an output current constant regardless of loading.

Differential Operational Amplifier A type of electronic amplifier that amplifies the difference between two voltages.

Digital-to-Analog Converter (DAC) A converter that converts digital data (usually binary) into an analog signal (current, voltage, or electric charge).

Electrical Instability An electrical condition that is generic to all physical systems that use the amplification of signals, power, or energy in various forms where the system can undergo a negative resistance state (dI/dV is negative), or electrical runaway. Electrical instability can lead to amplification, oscillation, or electrical failure.

Electrical Over-current (EOC) An electrical event, where the current magnitude exceeds the safe operating current of the electrical component leading to electronic system damage and failure.

Electrical Overstress (EOS) An electrical event, of either over-voltage or over-current, that leads to electrical component or electronic system damage and failure.

Electrical Over-voltage (EOV) An electrical event, where the voltage magnitude exceeds the safe operating voltage of the electrical component leading to electronic system damage and failure.

Electromagnetic Interference (EMI) An electromagnetic disturbance that affects an electrical circuit due to either electromagnetic induction or electromagnetic radiation emitted from an external source.

Electromagnetic Compatibility (EMC) A branch of electrical sciences which studies the unintentional generation, propagation, and reception of electromagnetic energy. Electromagnetic compatibility must address both the susceptibility of systems to electromagnetic interference and the propagation of electromagnetic noise.

Electronic Fuse (eFUSE) An electronic device which can be programmed to change electrical state.

Electrical Safe Operating Area (E-SOA) An electrical regime or state in current and voltage where an electronic component or system can operate without permanent degradation, latent damage, or failure.

Electrostatic Discharge (ESD) Electrostatic discharge (ESD) is a subclass of electrical overstress and may cause immediate device failure, permanent parameter shifts, and latent damage causing increased degradation rate.

Electrostatic Discharge (ESD) Power Clamp An electrostatic discharge circuit, which is used to provide current conduction between the power rails due to an ESD event, to provide a low current path and connectivity to the power rails.

Electrostatic Discharge (ESD) Protected Area (EPA) A manufacturing or assembly area which has the proper ESD control within the environment.

Electrostatic Shielding Shielding used in electronic systems to prevent the entry or penetration of electromagnetic noise.

Electrostatic Susceptibility   The sensitivity of a system to electromagnetic interference.

Equi-potential   A surface where all points on the surface are at the same electrical potential.

Error Amplifier   An amplifier with feedback unidirectional voltage control circuits where the sampled output voltage of the circuit under control is fed back and compared to a stable reference voltage.

ESD Control Program   A corporate program or process for addressing electrostatic discharge issues in manufacturing and handling in a corporation.

Feedback Loop   A connection typically between an output and input node of a circuit to provide positive or negative stability or sensing a signal.

Field Induced Charging   Charging process initiated on an object after placement within an electric field. This is also known as Charging by Induction.

Fuse   A circuit element to prevent over-current.

Gas Discharge Tube (GDT)   A gas discharge tube is an electrical overstress (EOS) device which is a switching device containing a gas between two electrodes. This high voltage device switches after electrical ionization, arc discharge of the gas. It is a bidirectional EOS element.

Guard Ring   A structure element contained within a semiconductor to collect, recombine, or current transport in or out of a device, circuit, or core region.

Human Body Model (HBM)   A test method whose source is a RC network with a 100 pF capacitor and 1500 ohm series resistor.

Human Metal Model (HMM)   A test method that applies an IEC 61000-4-2 pulse to a semiconductor chip; only external pins exposed to system level ports are tested. The source can be an ESD gun that satisfies the IEC 61000-4-2 standard.

IEC 61000-4-2 System Test   A system-level test method that applies a IEC 61000-4-2 pulse to a system for evaluation of system level robustness; the source can be an ESD gun that satisfies the IEC 61000-4-2 pulse waveform.

IEC 61000-4-5 Transient/Surge Test   A system-level test method that applies a IEC 61000-4-5 pulse to a system for evaluation of system level robustness to transients and surges.

Integrated Circuit   An electrical circuit constructed from semiconductor processing where different electrical components are integrated on the same substrate or wafer.

Ionization   A method to generate ions from atoms. Ionization techniques include both electrical as well as nuclear sources.

Latchup   A process electrical failure occurs in a semiconductor component or power system where a parasitic pnpn (also known as a silicon controlled rectifier, thyristor, or Schockley diode) undergoes a high current/low voltage state. Latchup can lead to thermal failure and system destruction.

Latent Failure Mechanism   A failure mechanism where the damage created deviates from the untested or virgin device or system. A latent failure can be a yield or reliability issue.

Low Dropout (LDO) Regulator   A DC linear voltage regulator which can operate with a very small input–output differential voltage.

Machine Model (MM)   A test method whose source is a capacitor network with a 200 pF capacitor.

Matching   A concept where two elements have identical spatial and electrical characteristics.

Metal oxide Varistor (MOV)   A metal oxide varistor, such as ZnO, is a variable resistor element that is used as an EOS voltage clamp protection device. At low current, it has a high resistance, which is reduced at high voltage and high current.

Mismatch   A concept where there is a difference between two elements either spatial, electrical, or thermal.

Multi-phase Regulator   A regulator with multiple phases (e.g., two-phase or three-phase network.

Negative Temperature Coefficient (NTC) Device   A device whose resistance decreases with increasing temperature.

Non-plated Through Hole (NPTH)   A through hole in a printed circuit board.

Polymer Protection Device   A conductive polymer device which has a low resistance and low capacitance which can be used as an electrical overstress (EOS) protection element.

Positive Temperature Coefficient (PTC) Device   A device whose resistance increases with temperature. As an electrical overstress protection device, it has a low resistance at low currents and high resistance at high currents.

Power Supply Rejection Ratio (PSRR)   The ratio of the change in supply voltage to the equivalent (differential) input voltage it produces in the op-amp, often expressed in decibels.

Printed Circuit Board (PCB)   A surface or board which is used in electronic systems to hold integrated circuits, single components, and wire trace interconnects.

Printed Wiring Board (PWB)   A surface or board which is used in electronic systems to hold integrated circuits, single components, and wire trace interconnects.

Pulse Width Modulation (PWM)   A modulation technique that conforms the width of the pulse, formally the pulse duration, based on modulator signal information.

Safe Operating Area (SOA)   An electrical regime or state in current and voltage where an electronic component or system can operate without permanent degradation, latent damage, or failure.

Silicon Controlled Rectifier (SCR)   A semiconductor device or component that can be used for electrostatic discharge (ESD) or electrical overstress (EOS) protection. This device is also known as a thyristor and pnpn device.

Static Electricity   Electrical charge generated from charging processes that are sustained and accumulated on an object.

Surface Mount Device (SMD)   A device which mounts directly to a printed circuit board without a through hole connection.

Surface Mount Technology (SMT)   A technology that allows for direct mounting of a device onto a printed circuit board.

Surface Resistivity   The resistance of a material on its surface (as opposed to a bulk resistivity).

System Level IEC 61000-4-2   A system level test that applies a pulse to a system using an ESD gun.

Switch Mode Power Supply Converter   An electronic power supply that incorporates a switching regulator to convert electrical power efficiently.

Thermal Instability   A thermal condition that is generic to all physical systems that use the amplification of signals, power, or energy in various forms where the system can undergo thermal amplification, or thermal runaway. Thermal instability can lead to amplification, oscillation, or thermal failure.

Thermal Safe Operating Area (T-SOA)   A state in current and voltage where an electronic component or system can operate beyond the electrical safe operating area (E-SOA) and below the region of thermal failure. In the T-SOA regime, permanent degradation, and latent damage can occur.

Thyristor Surge Protection Device (TSPD)   A semiconductor device or component that can be used for electrostatic discharge (ESD) or electrical overstress (EOS) protection. This device is also known as a silicon controlled rectifier (SCR) and pnpn device.

Transient Voltage Suppression (TVS) Device   A semiconductor device or component that can be used for electrostatic discharge (ESD) or electrical overstress (EOS) protection to address transient phenomena, electrical overvoltage (EOV).

Transmission Line Pulse (TLP)   A test method that applies a rectangular pulse to a component (10 ns rise and fall time; 100 ns plateau).

Very Fast Transmission Line Pulse (VF-TLP)   A test method that applies a rectangular pulse to a component (1 ns rise and fall time; 10 ns plateau).

Voltage Regulator   A circuit that established a fixed constant voltage.

Zener Diode   A semiconductor device or component that can be used for electrostatic discharge (ESD) or electrical overstress (EOS) protection. This device has a high breakdown voltage and usually is used in a breakdown mode of operation.

Index

ADCs *see* analog-to-digital converter (ADCs)
analog circuits
    Analog-digital converter (ADC), 255
    bandgap reference, 49, 255
    bandgap regulators, 36, 39, 52
    boost DC/DC converters, 46
    buck/boost converters, 46–8, 255
    buck converters, 46, 255
    comparators, 36, 39, 43, 52, 255
    current mirrors, 20, 36, 39, 43–5, 52, 77, 78, 256
    DC converters, 255
    differential operational amplifiers, 256
    low dropout (LDO), 75, 128–9, 197, 257
    operational amplifiers, 49
    phase lock loops (PLL), 50, 77
    receivers, 36–43, 52, 55, 56, 78–81, 90–92, 94, 101, 102, 104–6, 108, 109, 224–6, 246
    regulators, 20, 36, 39, 46, 52, 73–6, 197, 258, 259
    switches, 46, 47, 255, 257
    system clocks, 77
analog design
    active guard rings, 173, 187–90
    analog–digital (pre), 40, 182–4, 216
    analog domain, 77, 82–90, 92, 93, 96
    guard rings, 7, 8, 90, 171, 173, 182, 184–7, 189, 190, 211, 213, 233–5, 237, 239

mixed signal, 2, 49, 78, 82, 84–7, 89, 92, 96, 103, 116, 124, 130, 144, 171, 173, 199, 201, 202, 207, 208, 220, 223
analog–digital mixed signal design synthesis
    analog-to-digital guard rings, 90, 171
    breaker cells, 85
    digital to analog design ESD solutions, 81–3, 89, 97
    digital to analog signal lines, 90–92
    floorplanning, 116
    guard rings, 90, 171
    power domains, 85, 89
    power grid, 77
analog ESD power clamps
    bipolar ESD power clamps, 144–5
    CMOS ESD power clamps, 55, 120, 125, 140, 141, 143
    high voltage power clamps, 55, 62, 72, 120
    low voltage power clamps, 55, 62, 72, 130
analog layout
    array, 20, 22–4, 26, 29, 35
    capacitors, 22–6, 30
    common centroid design, 22–5, 29, 30, 36
    differential circuitry, 28
    diodes, 22, 24, 33–4
    resistors, 20, 22–30
    thermal lines, 19, 21, 22

ESD: Analog Circuits and Design, First Edition. Steven H. Voldman.
© 2015 John Wiley & Sons, Ltd. Published 2015 by John Wiley & Sons, Ltd.